WISDOM
OF THE TOOLS

What directs the world's work?

By Bill Merrill

ISBN 0-9773348-1-3

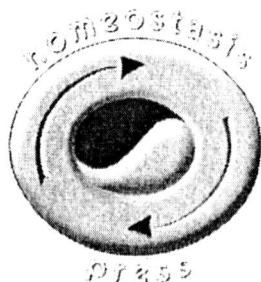

Published by Homeostasis Press, 315 SW Brent Drive,
John Day, OR 97845

DEDICATION

I gratefully dedicate this book to my mentors and colleagues in the 1960s Geography Department at the University of Oregon and to my brother Ted, who provided gentle spur and encouragement; and of course to my late wife, Jo, who, in addition to everything else, provided the constant backdrop of supreme humanity against which the inhumanities of the day show stark and clear.

FOREWORD

This book is an artifact excavated from the kitchen-midden of my files, datable in the era of 1968-'69. It evolved out of a growing conviction that far too little attention was being paid to the real ways in which burgeoning industrial energy was being controlled and directed.

Some potentially catastrophic 'unforeseen consequences' of our choices were edging into view. Among them (but by no means all):

1) *The supplies of readily accessible resources such as metals, fossil fuels, forests, soils and dam-able rivers being spent with reckless abandon.*

2) *Steady emergence of a global analog of a giant wheat field*: a mechanical-electronic monoculture increasingly vulnerable to 'unforeseen consequences.'

3) *Global warming*: measurements in the 1950's already showing that the ground work had been laid for accelerating changes to the gaseous mix of the atmosphere, which would inevitably alter the retention and distribution of solar energy—which is to say, climate.

4) *Forced evolution of new pathogens*: hospitals becoming hotbeds of antibiotic-resistant staphylococcus organisms. Malaria rebounding similarly from DDT assaults in the tropics.

5) *Relentless subsidization of private over public transport*: air pollution rising with the demands on finite fossil fuels. Cities forced into unsustainable patterns, designed for automobiles (see early Jane Jacobs).

6) *Degradation of real work*: accelerating capital attention on 'productivity,' which is to say, 'dis-employment.'

4

7) *Steady perversion of new science to weaponry*: CBW, aircraft, ancestors of Star Wars, etc., etc; etc.

Buckminster Fuller, in the lecture referred to in these pages, observed that, as a young man, he had been urged to get a job. However, he concluded that his proper course of action should be simply to set about doing work that he could see needed doing and hope for the best.

It was in this spirit that I wrote these pages; and it is in this spirit that now, in my dotage, I attempt to exhume them and set them free. I hope that this work, done a generation ago, might cast some light on how we have achieved our present perilous state, with once-unforeseen consequences besetting us on every side.

PREFACE

My physician-woodcarver brother has written an admirable foyer to this work. With his habitual clarity and immediacy, he shows how tools change their users, a dimension of our predicament largely neglected in this text. It was published in *The Country Journal*, 1974.

The Attitudes of Our Tools
By Ted Merrill

A major difference between us and other animals is that we make and use tools. Furthermore, to have and use a tool changes us, changes our relationship to the environment—and to each other.

When logging was done by lumberjacks, with two-man felling saws and teams of horses, a normal day's work resulted in a normal day's accomplishment in logs produced. Now, the same number of men with chain saws, tractors and trucks doing the normal day's work bring out a normal number of logs—which is several times more than in the old days.

The most important change that has taken place with the advent of chain saw and tractor is *attitude*—in the expectation, in definition of "normal" day's work achievement, and therefore in planning the work for tomorrow and next year. Thus people's *change in attitude* seems almost to reside in the tools themselves.

For several years I have done some woodcarving and small sculptures up to about 18 inches in length. Last summer, in order to do some larger pieces, I bought a chain saw and acquired some big elm logs. Armed only with hammer and chisel, I had thought of a "big" log as meaning maybe 1 x 3 feet; but with the chain saw in hand my relationship to the log changed: I had to hire a tractor to bring home the log for my first project. Though I still do some small-scale carving too, my attitude—my expectation and intention toward log and sculpture—was irreversibly changed by experience with my new tool. The chain saw's attitude toward a log—or a forest—is much different from that of a chisel or even an axe.

Last summer, hearing a roar and clatter down in Great Brook where it runs by our house, I went down to see what was happening. As I watched, a bucket loader moved up the channel, scooping gravel and rocks, some as big as a good-sized suitcase, out of the middle and dumping them on the bank. A couple of small dead elms near the stream had been cut down and moved up near the road. The little pothole where my daughter had caught a trout the month before no longer existed.

A bit later, the bucket loader broke down briefly and I was able to speak to the operator, who pointed out that he was merely doing what he was instructed to do. He referred me to his boss.

Fortunately the boss, Mr. Duranleau, was just arriving in time to receive my anguished protests and questions.

It turned out that in the wake of the June flood, federal funds had become available to the town to repair damage along the stream channel. The project in this area consisted of clearing out logs and trees which might, in a future flood, float down and jam up the several bridges along Great Brook, causing a danger of washing out the bridges.

Now, I am a bit of a fanatic about leaving things as they are, especially streams; and I think small, having never operated a bulldozer. So I asked why one wouldn't just take the logs out of the stream with a cable and winch, one at a time, or even get three or four men to do the smaller ones by hand. Mr. Duranleau suggested that it would be hard these days to find anybody who would want to do that sort of work. I moaned about the bucket loader gouging up the channel (which in this particular stretch of brook had been relatively unaffected by the flood), and about making wounds in the flesh of the earth which will take years to heal; about how it would take at least one or two more floods now to settle the rocks back into stable positions and thus re-establish a secure channel. Mr. Duranleau pointed out that he didn't wish to remove any more rocks or earth than necessary to get the machines into the brook, and that he, too, was interested in preserving some fishing holes. But he was doing what he had been hired to do, in the best way it should be done, to the best of his ability.

Eventually I cooled off. Great Brook is still there, and in our area it doesn't really look very different from the way it was a year ago. No doubt my daughter will catch another fish this summer. Perhaps I should buy a bucket loader, so that my attitude toward the earth will become more conventional and practical; for I think that once you own a back-hoe, the likelihood of your ever again digging a ditch one foot wide with an ordinary shovel becomes very small.

TABLE OF CONTENTS

Page

Chapter 1

WHO OWNS THE SHOVEL?

Nobody else will grant
like he said the volcano any
one of us does
sit upon, in quite such a tangible fashion
Charles Olson

In his moving film *Come Back Africa*, Lionel Rogosin conveys with crushing intensity the plight of a man from the South African bush, taught the rhythm of a shovel and sent down into the black and noisome bowels of the Great Dike of Johannesburg; there to work on a meaningless schedule at meaningless tasks, surrounded by swarms of others like himself. From those around him he draws his meager comfort, buries his fears and builds small meanings. For this man, the fears are close to the surface, the unknowable within arm's reach. His children may glimpse a larger view of the stream he is caught up in, and arm themselves with larger meanings. But the man with the shovel faces a lifetime torn between nostalgia for a familiar world whose paths his feet know, and the myriad messages he gets from mysterious authorities around him which tell him that his familiar world is dead, valueless and a trap; that salvation lies in the rhythm of the shovel.

Who really owns the shovel? Of what machine is it a moving part? For two million years the Community owned the shovel, and its handle was always warm from the palms of the fathers. Its rhythm was the rhythm of life, love and the seasons. The shovel turned up real earth in real mounds for purposes that everyone shared. The labor was not less—but it was life, not labor. The schism separating labor from life is new, not old.[1] Convention has it that domestication of plants broke the spell. Marshall McLuhan suggests the invention of the phonetic alphabet and moveable type. Whatever it was, it decreed the end of man's most successful social form: a small tribal non-literate community, which had brought him from the primates to his human estate and preserved the species for some 65,000 *Homo* generations. In the last fifteen of those generations a driven people from Western Europe has transformed mankind. With irresistible force they have destroyed the community and replaced it with the nation state; replaced the familiar shovels of wood and iron with strange ones of forged steel, aluminum and titanium; and substituted progress for life.

Who owns the shovel? Who composes the myriad messages from mysterious authorities that tell us that our old familiar lives without autos and antiperspirants are dead, valueless and a trap? How has it come about that the task of processing 50,000 human corpses and a million potential corpses swept up annually from the wreckage of automobiles has become simply another addition to the Gross National Product? How has it come about that courses of action clearly leading to eventual species suicide becomes identified with "hardheaded realism" and opponents to these courses are "dreamers" or "impractical idealists?" How has it happened that conventional wisdom has become, in a word, insane?

I sincerely believe that we do not know. I believe that over the last two centuries industrial people have built such a dense screen of illusion between our flesh-and-blood selves and the workings of our technologies that public and private discourse has become one long litany of *non sequitur*. The

---

[1] In a small, tribal non-literate society, the distinctions by which modern heterogeneous society is ordered could hardly apply. Our conventional dichotomies of work/play or sacred/secular could have had no meaning. All activity took place within a seamless web of legitimacy. The contention here is not that life was physically easier, but it was integrated, undivided.

young sense the monstrosity of this disjunction, and rebel against it. That they sense it at all is a measure both of their sensitivity and the incompleteness of their education. But since their literacy is the literacy of the industrial cultural *non sequitur*, their rebellions reflect its limitations. Thus students obstruct Dow Chemical's recruitment on campuses, but drive Fords and Chevrolets. They attempt to root out military research projects, but leave departments of advertising and public relations in peace. This is, of course, not surprising. It would be extraordinary indeed if one whose entire experience with language was limited to the prose and poetry of Lewis Carroll, were thereby enabled analytically to criticize Lewis Carroll as an author.

Industrial energy has been and is currently thoroughly misunderstood. The multitude of unwanted and unforeseen consequences of its development and deployment provide ample testimony. Neither vernacular nor expert understanding has served the ends of anticipation of the consequences, either physical-biological or social, of the exponential increase in amount and forms of energy released upon the world by industrial tools and techniques. The chronicles of disaster that form a kind of substratum under contemporary literature have this to say above all else: the social and mechanical precepts of our own day are failing us.

A search for new conceptual frameworks that can illuminate what is going on may, perhaps must, begin with a calculated naivete. The most deeply rooted and widely accepted precepts must be held suspect, to be called back only with empirical justification. The conventions of economics do not explain the transformation and distribution of industrial goods. The conventions of politics do not explain the role of industrial agencies as political entities, nor the impact of industrial energy on social-political institutions. To declare useless such stately edifices as have been built by the economists and political scientists is not my purpose. They obviously have enormous importance. Since they do not explain, then what is their purpose? If the conventions of the market and nation state are leading us into catastrophe, why are they still the stuff of survey courses in how the world works? Any new view must subsume and explain the persistence of these notions. A functioning conceptual model of the industrial realities must encompass institutionalized illusions as well as the nature of tool systems.

Industrial tooling has progressed far enough that certain perspectives can now be attained that were quite unattainable only one or two decades ago. Some of these may be derived from the world-wide data collection and compilation by the United Nations, in itself a sophisticated industrial task. One of the most fascinating of these relates to the distribution of energy consumption. Though the actual degree of concentration and asymmetry is masked by any figures aggregated by nations, the picture of energy consumption that does come through is highly instructive. Measured in equivalent units (kilograms of coal equivalent), and omitting wood, peat and dung, the energy available to and consumed by the population of highly industrialized nations is roughly 1000 times that of the peoples of the least industrialized.[2]

If the data allowed finer discrimination than whole national populations, the concentration would be revealed to be many times greater than that. Certainly the 10,000 units per capita consumed in 1966 by the people in the United States (as opposed to 8 units per capita in Chad and Yemen) are not uniformly distributed. Consider, for example, the enormous energy at the disposal of the controllers of a highly automated petroleum refinery; or the crew of a B52 flying sorties over Vietnam; or the Atomic Energy Commission.

Another kind of distortion is inherent in the United Nations data, which masks concentration. The energy at the disposal of IBM-France (a subsidiary of IBM) shows up as part of the French supply. The principle offices of IBM (the parent corporation and therefore the source of ultimate

---

[2] UNESCO, World Energy Consumption.

direction over the energy) are located in the United States. This understates the energy under the direction of agencies in the United States.

Such differences in energy availability are something wholly outside the scope of previous human experience. It is worth noting, perhaps, that the much smaller advantages held by Europeans in the 19th century were sufficient to be completely decisive in conflict with non-Europeans. By the beginning years of the 20th century, it was clear that the fate of any non-industrial social form was sealed, if challenged by a people with access to industrial tools.[3]

The enormity of the challenge in amount of energy at the disposal of human agencies and the peculiar distribution of that energy lie at the center of the quandary. What is this phenomenon we call the "industrial revolution?" Is it a continuum with what preceded it, or does it mark some sort of quantum break with the past?

The chronicles of disaster suggest the latter. Without caviling over the precise historical moment of the break, what it means to Man as a species is becoming all too clear. The present presence on earth of *Homo sapiens* is testimony that his life ways have added up to a net survival advantage. But never before could a single set of human acts, within the life span of a single man and committed within a few or a few hundred square miles, transform the entire earth environment. It is this aspect of the industrial revolution that is giving rise to my conviction that we are truly confronting something that has never existed before in any form: the physical possibility that purposeful acts can set in motion

---

[3] Picking fixed dates for breaks in historical streams is perhaps more fun than definitive. In this case, 1870 has certain attractions. Imperial China was helpless, her customs and port cities in the hands of foreigners. Japan was committing herself to become European in order to remain Japanese. South Asian peoples were under European flags. Trans-continental railways were putting the final stranglehold on the relict peoples of North America. And Europeans were crouched on the starting line to overrun Africa.

such massive and rapid changes that there is simply no chance that the web of biotic material can continue to absorb them. Being suddenly able to alter the whole at once, we begin to plan in world-comprehensive terms. But the plans lack a crucial ingredient: concern for species survival. Previously, cultural groups could die and mankind would go on. Groups with certain dysfunctional life ways would simply disappear, and others more successful would occupy their space. The industrial revolution has placed all at the mercy of any. Industrial man has undertaken to plan, and has perhaps elaborated the power to do so. But the plans are insane, if survival of the species is thoughtlessly jeopardized by the plans.

In the following chapter, I attempt to dispel any doubts that the human species is indeed in deadly and imminent peril.

Chapter II

DESIGN FOR DEATH

*I think only Americans could observe their
world today and be optimistic about it.*
                                        Carl O. Sauer

A library of doom graces my desk. Inside its colorful volumes are perhaps the most grisly treatises to appear at any time or place. Their net effect if read consecutively might be compared to the scrutiny of a 5000 page field manual for Buchenwald. These books have appeared since World War II. They are the books of the time of nuclear weapons, DDT and 3% per year population increase. They are apocalyptic books, books of profound pessimism concerning the future of the whole human species. Similar writings of the past have dealt with localities, with problems limited to particular places and conditions. Not so these. The whole world and the whole human species are the subjects. Yesterday's concern with the silting of rivers has become today's admission of chemical toxicity of the waters of the ocean margin. Yesterday's city smoke has become today's change in the gaseous mix in the entire lower atmosphere, endangering the delicate balance of solar energy gain and loss on which all living things depend. Yesterday's concern for exposure to X-ray emissions in shoe stores and dentists' offices has given way before the recognition that each man in the world is his own source of dangerous radiation, squirreling away in his bones the fruit of nuclear testing and power generation.

One who would take refuge from these dismal thoughts in books of refutation searches almost in vain. There are optimists among us, but their arguments rarely meet the arguments of doom on the same ground. Pierre Teilhard de

Chardin offers hope,[4] but it is the hope of his Jesuit faith, not of his science. Stuart Chase offers hope, but reads like well-earned wishful thinking.[5] One may ignore the dangers of technology and concentrate attention on its wonders. After all, all the evidence is that *Homo sapiens* will not disappear—he hasn't yet. By the same logic, of course, all the evidence is that any automobile that has not yet wrecked, won't. Its entire history testifies to its future immortality. The measure of this logic can be found in auto wrecking yards in the immediate vicinity of almost everyone.

Routine optimism is implicit in the *assumption* that nothing very terrible will happen. Typical is the initial material published by the American Academy of Arts and Sciences Commission on the Year 2000, under the chairmanship and editorship of Daniel Bell.[6] Environmental pollution and the bomb are almost completely ignored. So is hunger, although limited obeisance is done to population increases. It is almost as if the Commission had decided *a priori* that if they admitted these phenomena to the future, it

---

[4]*The Phenomenon of Man.* Harper and Row, New York and Evanston, First Harper Torchbook edition, 1961, pp 285-290. In part, of course, Teilhard's apparent optimism simply reflects his choice of what to write about; his decision to deal with the peaks of human possibilities and let others deal with the abysses.

[5] *The Most Probable World.* Harper & Row, Pelican Books, 1969. Particularly disturbing is Chase's use of the idea of the trend line. He takes comfort from Platt's observations concerning the tendency of exponentially rising curves to turn into S curves. The reality of this is hardly deniable. Who could argue that the present geometric increase in the world population is in fact going to continue indefinitely? Or that the rate of increase in DDT in the oceans is temporary? To take comfort from these conclusions is something else. As an infinitude of such curves have risen together, so they may decline together. Is there any assurance that the decline will be self-limiting? It is certainly within the possibilities of the situation that we may discover experimentally that some wonder of technology has effectively eliminated us as a species, as is occurring so dramatically today to carnivorous birds. This will adequately dispose of all rising curves that threaten us.

[6] *Toward the Year 2000: Work in Progress.* Beacon Press, Boston, 1969.

would impair the fit of their predictive models. The most substantial comment on environmental pollution in the entire volume was made by Harvey S. Perloff in a paper on the future of the urban scene: "If certain of our present urban trends were to continue, we would have some extremely serious problems on our hands in the year 2000. Water and air would be dangerously polluted. An increase in pollution at rates now characterizing some of our bigger cities would make relatively pure air and water among the scarcest and costliest of all natural resources."[7] This is optimism by default, or the ultimate in safe use of the trend. The trend of a balloon is to get larger as it is being inflated, but only for a while.

William Vogt began this new generation of books of doom immediately after World War II, with *Road to Survival*.[8] Its theme combined concern for destruction of landscape, viewed as resource, with concern for the proliferation of people. It questioned the morality of outsiders investing in the survival of larger and larger numbers of individuals, while ignoring the consequences for populations. It raised basic questions about the virtue and survival chances of high-energy societies.

Six years later, in 1954, Harrison Brown added a book developing the same themes, *The Challenge of Man's Future*.[9] His work differs from Vogt's in emphasis. It is more specifically focused on the energy requirements of industrialization, and seriously discusses the likelihood of the early demise of machine-oriented culture. He also made explicit the well-nigh irresistible pressures industrial society places upon agrarian peoples to become industrial; that is, to join the march to oblivion. His book is one of measured pessimism, suggesting that Man's most likely future includes

---

[7] *Ibid.,* p. 157.
[8] William Sloane Associates, Inc., New York, 1948.
[9] The Viking Press, New York, 1954.

a massive Malthusian die-off and a permanent reversion to agrarian life.[10]

Marston Bates, a biologist, addressed himself to the neo-Malthusian problem, concentrating on circumstances of population. *The Prevalence of People*[11] appeared in 1955, nearly simultaneously with Brown's book. Taken together, these books represent well the state of the art of foreboding in the mid-1950s.

After 1955, landmark foreboding became specialized. Of all books of the genre, the most important appeared in 1962, Rachel Carson's *Silent Spring*.[12] An appalling chronicle of environmental poisoning, it was the first post-war book of warning to stimulate Congressional investigations into the state of the world's biotic environment. It was the first book of its kind seriously to threaten the freedom of action of the nation's major chemical polluters and promoters of pollution. Whereas previous books had been little noticed outside of academe and the church (some clerics were edgy about Vogt's outspoken and irreverent advocacy of birth control), Carson found herself facing the unified front of the petrochemical companies and their patrons and satellites, from the university chemists to the Department of Agriculture.[13] It is probable that public tranquillity will never

---

[10] *Ibid.,* p. 226.

[11] Charles Scribner's Sons, New York, 1955. "While this manuscript was in its final stages, a book appeared written by Harrison Brown ...Mr. Brown is primarily looking to the future, while I have been trying to understand how we arrived at the present." P. 247

[12] Fawcett Publications, Inc., Greenwich, Conn., 1962.

In an Address at the University of Oregon in August 1969, Garrett Hardin rates *Silent Spring,* with the discovery of DNA, as one of the two most important events of 20th century biology. If in the end we escape the peril of which she warned us so eloquently, the debt owing to this gentle lady will be of a unique order.

[13] Shortly after excerpts from Miss Carson's book were printed in the *New Yorker* in June 1962, she became the center of a proper uproar. Lines were drawn, and combatants took positions which they essentially maintain today. The petrochemical, food, and other industries with

quite recover from *Silent Spring*. Time passes, but as the book might be forgotten, its predictions appear in the news in

---

financial interests in the pesticide market took the position that modern living would not be possible without chemical pesticides. Interesting refinements include the notion that creation of hostile public opinion toward chemical pesticides would discourage industry research, hence perpetuating the use of the present types, whatever their shortcomings. (*New York Times,* Nov. 18, 1962, p2.) Miss Carson, spokesmen for the industry maintained, had painted a distorted picture by not giving equal time in her book to the benefits enjoyed from the use of the hard pesticides she attacked. Typical of these comments, and of the common cause made by industry and university chemists, was the report of the annual meeting of the American Chemical Society appearing in the *NYT* of Sept. 13, 1962, p. 34. Conservation groups and large numbers of private citizens aligned themselves predictably with Miss Carson.

Government, in the U.S. and abroad, vacillated. The President's Science Advisory Committee issued a report in May 1963, which acknowledged the seriousness of the situation and recommended strong measures. After the issuance of this report, the Department of Agriculture reportedly opposed and helped kill a bill in Congress requiring that pesticides toxic to wildlife be so labeled. (*NYT,* Nov. 10, 1963, p. 50) Meanwhile, a Food and Agricultural Organization conference had urged U.N. member nations to take strong measures against pesticide hazards, but not so strong as to impair international trade in food. (*NYT* Nov. 13, p76; Nov. 16, p. 10, 1962.

Domestically, the Pesticide Committee of the Federal Council for Science and Technology arrived at the curious conclusion that 1) hazards of pesticides could not be estimated because of inadequate knowledge; and 2) risk must be scaled against benefit. (*NYT,* Nov. 18, 1962, p 41.)

Five years later, personnel of a local U.S. Soil Conservation Service office attempted to dissuade my colleague and myself from including *Silent Spring* in the curriculum of a conservation workshop for teachers. And on Oct. 9, 1968, the *NYT* said, reporting on a Department of Agriculture study of pesticide use on food crops. "Pesticides and other chemicals are used on food crops under Federal controls designed to keep foods free of unsafe, high-level chemical residues."

One might hope that the contending forces would close ranks following a *NYT* story of May 6, 1969: "Dr. Goran Lofroth, a Swedish toxicologist and expert in environmental poisoning, reported that breast fed children around the world average twice the daily intake of DDT compounds recommended by the World Health Organization as a maximum, putting these children in the range of exposure inducing biochemical changes in laboratory animals."

21

the form of additional species of fauna falling victim to DDT, or shipments of food confiscated by order of the Food and Drug Administration for containing too much pesticide. Within 6 months of the time of this writing, to the list of victims have been added the east coast peregrine falcon; the California brown pelican; the San Francisco Bay shrimp; the crab population off the Golden Gate; the newly established population of coho salmon in Lake Michigan (declared unfit for human consumption by the FDA).[14]

A legacy of Rachel Carson's time, and perhaps of her book, is a tendency of the foreboding books thereafter to take the face of imminent doom almost for granted (albeit reviewing its elements), and to focus attention in greater depth on one or another aspect of its causes, effects or remedies. *Silent Spring* was followed in 1963 by Barry Commoner's *Science and Survival*.[15] Commoner's central concern turned to the mechanisms of scientific failures through success, as it were. How is it, he wondered, that scientists' work escapes scientific controls and serves to threaten our species with the tools and turmoil it creates? He is looking for ways of reordering science and scientists in order that this once powerful expertise may be held to the interests of species survival.[16] Commoner's book is more pointedly political than its predecessors, in that he makes a solid effort to define the areas of information and value

---

[14] "The Biology of Pesticides," *Cry California,* Summer, 1969, vol. 4, no. 3, pp. 2-10.

In addition to listing victims, these authors point out that there are approximately 1 billion pounds of DDT now loose in the environment. This will, with no more added, exhibit a half-life of about ten years, continuing to concentrate up the food chains. On April 23, 1969, the *NYT* reported seizure by the FDA to that date of 34,000 pounds of coho salmon from Lake Michigan, ranging in DDT content from 13 to 19 parts per million. The Food and Drug Administration had not previously provided for any DDT content in fish, but set interim limits of 5 parts per million.

[15] New York, The Viking Press, 1967 (first published 1963).

[16] *Ibid.*, Chap. 2.

judgments necessary to the perpetuation of high energy society; and to draw clear lines between what can and cannot be served by scientific expertise. [17]

Two more examples from this grim library illustrate distinct types. Paddock and Paddock, *Famine—1975!*,[18] focuses on the population explosion *vis a vis* food supply; argues for the inevitability of a catastrophic human die-off from hunger beginning in the middle of the 1970s; and then, assuming this as inescapable, looks for policy with which to meet the crisis.[19] The distinctive aspect here is that the policy sought is not one by which the catastrophe might be avoided. The Paddocks see that chance as already gone. The policy question posed is rather, what do we do when it happens? The Paddocks' book is to Vogt's book as an epitaph for Man would be to *Silent Spring*.

Nigel Calder's anthology of present and future weapons, *Unless Peace Comes*,[20] completes this particular lineage of writings. Whereas the other books have dealt with the inadvertent destruction that the uses of enormous energy can inflict on Man, this one, through the views of sixteen experts in various fields of weapons technology, show how we are designing that energy in various ways to bring about the same end on purpose. In some ways the most profoundly pessimistic book of the lot, it leaves one with the conviction that *any* breakthrough into new technical capabilities will be immediately subsumed into the arsenal for war, no matter what or by whom the finding. It would fit well as the technical supplement to that astonishing Galbraithian satire, *Report from Iron Mountain*.[21]

Famine, war, disease and death are not new, either as literary themes or as events. What is new is, first, the scale.

---

[17] *Ibid.*, Chap. 6.
[18] Little, Brown and Company, Boston and Toronto, 1967.
[19] Chapter 9 is devoted to the notion of "triage" in dealing with the inevitable "age of famines."
[20] Viking Press, New York, 1968.
[21] Anonymous, The Dial Press Inc., New York, 1967.

These authors consider that it is not a city's or a nation's fate at stake, but that of *Homo sapiens*. Energy is available for an infinitude of kinds of work in the world described by these authors. Its present magnitude was inconceivable to the industrial public of one generation ago, and is increasing exponentially. Its impact is commensurate with its magnitude. Autos pollute the world's atmosphere and oceans with lead. Agricultural pesticides poison the entire ocean spaces. The logic of the energy scale would make it simple justice for the Congolese to have a vote on the question of additives used in American gasoline, because they too are imbibing the effluent.

A corollary of this world scale is an undertone in these writings, evidenced in part by the remarkable representation of biologists,[22] of a search for, and demand for, an ecological view of Man's existence. They are asking, explicitly or implicitly, that we stop trying to defeat our companion organisms and rejoin them as part of the total biota.

The aggregate message views Man as being in imminent jeopardy from his own technology through any or all of three main streams of events, any one of which threatens him as a species directly and by threatening necessary ecologic links to the larger biotic world: the inadvertent poisoning of the biotic environment; the proliferation of people far beyond the capacity of existing physical and societal systems to absorb, precipitating famine so widespread and intense as to bring about the collapse not only of the hungry societies, but the rich ones as well; or war with any of a wide spectrum of weapons of mass destruction, including nuclear, chemical, biological, or any of a host of promising developments which might be called "environmental" weapons of which one of the most intriguing is the creation of holes in the

---

[22] Vogt, Bates, Carson, and Commoner are all biologists.

ozone layer high in the atmosphere, which would permit the ultra-violet solar radiation to eliminate life beneath.[23]

Each of these in its own way is the result of the application of industrial-scientific techniques and energy to immediate situations without consideration of long-term consequences. The amount of energy and sophistication of application is so great and increasing so rapidly that the present seems to be overrunning the future. Situations become irreversible before they are perceived. The Russian Roulette mechanism of this is perhaps easiest illustrated by DDT, though the principle operates in any situation bringing about change that disseminates throughout the world (which is what industrial-scientific technology does).

Figures available to Rachel Carson in 1962 showed the average American having stored DDT through dietary sources to the amount of 5 to 7 parts per million.[24] Paul Ehrlich reports in 1968 that this has reached 11 ppm. [25] Since we have no way of knowing what the ultimate effects of DDT are on ourselves, any more than we anticipated that the pelicans and falcons would be rendered unable to produce viable eggs, it is possible that 10 ppm is the critical number for people. The only way we are likely to find out is by passing the threshold. Then it will be too late. And stopping production and use does not hold the contamination to present levels. All DDT now loose in the environment will continue to concentrate upwards in the food chain. The contamination of our own bodies, now about 3 times that permitted by the FDA in marketed fish, will undoubtedly continue to rise for several years.

But this example is only the clearest and most familiar, not necessarily the most ominous. Consider the same

---

[23] Calder, *op. cit.*. p.191. Gordon J. F. McDonald, associate director of the Institute of Geophysics and Planetary Physics at UCLA, has contributed a chapter to Calder's book entitled "How to wreck the environment (Geophysical warfare)."
[24] Carson, *op. cit.,* p.30.
[25] *The Population Bomb,* Ballantine Books, New York, 1968.

principal applied to phytoplankton. We depend for most of our oxygen supply on the photosynthesis carried on by these minute plants of the sea. Industrial processes are currently using the oceans as the direct or indirect dumping ground for hundreds of thousands of chemical substances, many of which are specific plant toxins. When the *New York Times* reports the oceans toxic to phytoplankton it will also be too late. Unfortunately, such a story in the *New York Times* would be quite by accident, since the *Times* does not cover phytoplankton routinely, any more than it covers falcon and pelican populations routinely.

Industrial society as now constituted cannot continue. At best it can commit somewhat selective suicide; at next best, total *homi*cide. *This is now being denied, but not refuted.* Technology seems to be guided by a kind of automatic pilot to oblivion. The task laid out for us by the Rachel Carsons and Harrison Browns is that of discovering ways of regaining control over our own machinery. Conversations over many years with students, the rhetoric of the New Left, and the anguish of the well-oiled residents of the shores of the Santa Barbara Channel all point to emerging traces of a healthy though tardy fear of technology itself. If that fear continues to grow, it may provide a climate wherein serious search for means of control and rational direction may be supported, but rational control awaits understanding. Rachel Carson put the question that may no longer be left as rhetorical. She sees future historians asking, "How could intelligent beings seek to control a few unwanted species by a method that contaminated the entire environment and brought the threat of disease and death even to their own kind?" [26] The spirit of this question can be extended to the entire range of industrial energy applications, and must be answered. That we are indeed behaving in this fashion is no longer a question. How is that we do this? Foregoing this

---

[26] Carson, *op. cit.*, p. 19.

question will mean foregoing even the satisfaction of knowing what happened to us.

Policy, public or private, which fails to incorporate provision for species survival is insane. Patriotic, virtuous, realistic, democratic, humane? Perhaps. But insane. Miss Carson is asking an historical question: how is that we have come to do what we do with our technology? If species survival is to find its way into the value system governing the use of industrial energy, that system must be discovered in the historical development of the Industrial Revolution.

Chapter III

PANDORA'S TOOLS

> *God,* means *are funny things!*
> *And* ends?
> ...*They also need tied up,*
> *But have fast legs.*
> Ronald H. Bayes

The ideas of Buckminster Fuller have been too important to me for relegation to the bottom of the page. His analysis of the "industrial equation" and his definition of the difference between "craft" and "industrial" tools constitute the point of departure for the next step in this discussion. Fuller's distinction makes sense to me, it seems eminently useful, and henceforth the terms "craft tool" and "industrial tool" will follow his usage.[27]

Fuller's distinction between industrial and craft tools is sharp. An industrial tool is one that cannot be made by "one man alone, naked in the wilderness," no mater how much he knows. Craft tools could be so made. Industrial tools augment the capability of other existing tools, exhibiting thereby a kind of multiplying effect on the total capability. Craft tools contain their whole meaning within themselves, and thereby only add to, and do not multiply, such capability.

The principle might be illustrated by the canoe and paddle, and the ball and race of a bearing. The canoe can exist without the paddle, and the paddle augments the capability of the canoe-and-paddle viewed as a system. Thus the paddle may be thought of as an industrial tool in relation to the canoe. This is about as far as the canoe-and-paddle can

---

[27] Buckminster Fuller, "Terminal Lecture," Visiting Artist Seminar, Summer Session, University of Oregon, 1962. Lecture is on tape at the University of Oregon Library.

be forced into the industrial category. One more canoe-and-paddle adds only its own transportation capability to the system. The possibilities of the canoe-and-paddle becoming a component of a larger system are quite remote. The canoe-and-paddle taken together compose, therefore, almost exclusively a craft tool. In the case of the ball bearing, on the other hand, its only function is as a component in some larger system of tools. It may augment the capabilities of any of an infinitude of tool complexes which happen to include turning metal shafts, but it has no use by itself. The ball bearing is, therefore, a pure industrial tool.

The industrial tool is not exclusively modern, nor the craft tool exclusively ancient. Early industrial tools only became highly visible when they began to agglomerate into powerful systems, illustrated by Fuller in terms of the great ships and the means of navigating them at will on the open oceans. Hosts of essentially industrial ideas had to exist before such elaborate systems could appear.[28]

Of supreme importance to the whole mechanics of the industrial phenomenon is the self-augmenting property of industrial tools. Granting the basic proposition, one is then propelled in curious directions. If a property of industrial tools is that they increase the efficiency or allow the utilization of higher energy levels through existing tool complexes, as the ball bearing or a new stress-resistant metal; then this higher energy level can be used to make still more tools that were not previously possible. Thus industrial tools can be conceived always to include, at any stage of the art, *tools to help make tools*. Fuller illustrates this by using again the early ships. Inadequately rigged, they would leave their European yards. At one stop they would acquire

---

[28] Industrial tools are easier to find in the kits of the ancients than are craft tools in my own house. Cordage, pulleys, levers, even language, have an industrial flavor. Even the various wooden artifacts around me were shaped with hardened steel tools, or glued with the aid of hydraulic presses. I suppose that the closest things to pure craft artifacts in my house are low-fired ceramics.

superior masts, adding them to their construction and taking additional ones as cargo bound for home. This would be repeated with cordage, sails, sheathing and other components. When the ship arrived home again it was not only a completed tool, but its cargo increased the capability of home shipbuilders to build superior ships. This simple and direct principle could be reiterated with iron and railroads. At later stages of sophistication, the relationships are more subtle but more powerful. An extraordinary complex of tools and knowledge, of which I have only the haziest notion, had to precede the control and fabrication of titanium. Once the existing tools provided this capability, however, access to this metal permitted a whole new round of possibilities. Tools-to-make-tools has taken on a staggering dimension in the cosmology of electronics, transistors, integrated circuits, computers, and all that these make possible.

Extending the logic of tools-to-make-tools produces a crucial conceptual notion or image: a network of tools whose functions interlock, growing on its edges and decaying in the center. As ever-higher energy levels and ever-greater sophistication are available on the experimental edge, new tools adapted to even higher energy levels are born. Obsolescence gradually overtakes the older components of the system and they disappear. Inventive effort is strongly biased toward tools capable of higher and higher energy expression. Thus the tool system guarantees that more and more kinds of tasks can be undertaken and that greater and greater energy can be delivered through the tools in the performance of any given task.

Another aspect to this picture is the logic of the *kinds* of tools generated. Two factors are always present on the experimental edge. One is what the inventor would subjectively like to invent—that is, the societal judgment of what kinds of tasks need doing—and the corollary of what kinds of tools are needed. The other factor is the existing family of tools, with their finite and qualitatively limited capabilities. Only those tools will appear in the next generation that the present family of tools makes possible. If conflict arises, it is obvious which will win. A new generation of tools will grow on the old edge, inventors will invent what is technologically feasible, and subjective social values will take potluck. Items from the menu of this social potluck continually surround us. The chaotic state of passenger movement in this country might be considered to serve the social values of our bitterest enemies, perhaps, but hardly of ourselves. (Teaching airport traffic controllers to love the supersonic transport may be the next major challenge.) Only a logical contortionist could maintain that the public has "asked for" the endless redundancy of non-goods whose prices include the cost of the merciless barrage of non-information necessary to its sale. Phony product differentiation, deceptive packaging and pricing, concealment of the cost of credit, enormously expensive propaganda, these are all goods sold, and at the same time tools used, nominally in the public interest. But in the universe of industrial weaponry, the primacy of the tools reaches the greatest heights of absurdity, when viewed from the interest of real people, with blood, bones and children.

Let us grant, for the sake of argument, the utter necessity for a "balance of terror" in weaponry, with its built-in guarantees of escalation. Public polemics concerning its form provide a crystalline example of the difference between the subjective needs of society and the compulsion of tools to whelp tools. America's current crop of weaponeers, public and private, hold that national security demands that all possible weapons systems be developed.[29] After all, the capability of industrial peoples is so nearly alike that the slightest lapse of diligence means that potential enemies can achieve a kind of weapon which ours will not "deter." The clear implication is that gas is only deterred by gas, nuclear

---

[29] It is worth noting, perhaps, that there is no visible lobby for spears, bows or swords. Body armor, offering possibilities for research, has lived on.

weapons by nuclear weapons, biological weapons by biological weapons, etc.[30] Each argument is subject to endless refinements. Each variation must be countered by the same variation. ICBM's deter ICBM's and MIRV's deter MIRV's. Submarine-launched missiles deter submarine-launched missiles. "First strike" capabilities deter "first strike" capabilities. Anthrax deters anthrax. Plague deters plague. Rift Valley fever deters Rift Valley fever. Nerve gas deters nerve gas. This is the perfect argument to justify unrestrained growth on the edge of the existing system of tools, but requires no particular expertise to detect the intellectual flatulence. Why do not nuclear weapons deter biological weapons? When weaponry reaches overkill status, why do doomsday devices need to multiply? Seymour Hersh described the policy defense against such questions as follows:

> Military critics of this position theorize that the United States would seriously weaken the deterrent to a biological attack by announcing a policy of nuclear retaliation. These critics point out that some of the possible warfare diseases take three or four days to break out. Would the United States, they ask, be willing to unleash nuclear missiles four days after a BW attack and tell the world it was "retaliating?" How would the United States be confident enough it had the right aggressor to order such an attack? Because of these and similar questions, proponents of biological warfare argue that a rigid policy of nuclear retaliation in case of BW could quickly be challenged by an enemy who would, in effect, dare America to counter germs with nuclear bombs [31]

---

[30] Defense Secretary Laird enunciated an official form of this argument as reported in an Associated Press dispatch appearing in the *Portland Oregonian,* July 29, 1969, p 4: "Secretary of Defense Melvin Laird said Monday the United States must stock chemical and biological agents as a deterrent against that type of attack by other nations against this country. 'That deterrent is important if we want to see that these gases … are never used in our time,' the defense chief said."
[31] *Chemical and Biological Warfare,* Doubleday & Company, Inc., Garden City New York, 1969, pp. 31-32.

The curious notion that BW retaliation against an unknown enemy would be possible, while nuclear retaliation would not; and the suggestion that a four-day delay would inhibit the willingness to release one kind of attack but not another; obviously have nothing subjective to do with national security. They are rationalizations for the ultimate value of technology: the continued elaboration of everything that the presently tooled-up busyness makes feasible; the efficient use of all the presently assembled teams of specialist experts; all the plant and laboratory capacity; all the teams of public relations and lobby folk; all the political clout and organized access to funds; all the professional-technical commitments of all the personnel now or prospectively busy. Tools grow on tools. The present tool kit, in and of itself, provides the necessary legitimacy for the next generation of tools, regardless of the human values of the work they are capable of. Hersh reports that at the time of his writing (1967) about forty scientists were working for the army at Fort Detrick on the problem of developing mutant strains of pneumonic plague and other diseases *which would be resistant to all known medication.*[32] Technology's response to success in such a voyage into the Dark Ages is wholly predictable: a demand for research funds to provide defensive capability against such weapons. We must have new medicines before the Russians do.

The military is demanding an anti-ballistic missile system for which the public may pick its own estimate of price—with a low bid of $5 (or is it now $7?) billion. The rationale is defense against the Soviet offensive capability. In CBW, Soviet superiority is argued as a rationale for an offensive capability, but defense in the present state of the art offers little opportunity for research and development. There are, therefore, not even enough gas masks now available to civilians to protect the employees of the Department of Defense, nor is there an adequate stock of antibiotics and gas

---

[32] *Ibid.,* p. 64.

33

antidotes. Nor have there been any public pleas for money. [33] Thus is intelligence brought to bear on the public weal.

A self-augmenting tool system, then, exerts strong pressure on the industrial society to find ways of learning to desire what is technologically feasible at the moment. This will at least supplement, and at most supplant, the notion that tools are devised to perform the tasks that the society, on non-technical grounds, concludes it wants. A case can be made, then, for technological dominance of values in industrial society in a fashion quite new to human experience. No craft tool system can so dominate its users. It is theoretically possible to decide on a task and invent a craft tool to aid in its performance, independently of the existing tools. Craft tools can be made by one man "naked in the wilderness." Not so in industrial society. Given our present family of tools, we must learn to crave the moon, or "…to stop worrying and love the bomb."

Learning to crave the moon might not be such a formidable task if we had a generation or so to build up some mythology about it. Unfortunately, it seems that having reached the moon, only a handful will get to see it close up before we are asked to junk it in favor of Mars and/or the newest undersea passenger liner to the bottom of the Atlantic Ocean. We found it easy to want antibiotics, but we are having more trouble really enjoying staphylococcus infections. Autos are fine, but smog is harder to develop a taste for than beet greens. The generation of new tools, each of which we must find it in our hearts to admire as progress; and the effects of its use, which we must learn to define as improved standard of living, no matter what the effects are; follow each other at a continually accelerating pace. And this is the other critical aspect of the mechanics of industrialization.

When tools-that-make-tools came to be the preoc-cupation of Europeans, the mechanical aspects of their

---

[33] *Ibid.,* p. 34.

34

societies set off on a new rate of accelerating change, different from the rate of change in other components of the culture. Most of our subjective values concerning good and bad, ought and ought not, accumulate gradually over our lifetimes, and we pass them on as intact as we can to our children. Tool generations have no fixed duration, as human generations do. Industrial tools follow on the heels of their predecessors at an accelerating rate, while changes in values remain rather firmly tied to the biologically decreed generation span. The "generation gap" may be wide or narrow, but the notion enjoying currency with the kids that the age of 30 is the threshold between the trustworthy and the mossbacks is solidly founded in human biology.

The tools of modern technology, then, appear to have three facets not commonly examined. First, we should find that they have tended to develop in a single matrix, a network growing on its edge and deliquescing in its progressively obsolescent center. Second, the tools are determining their own successors. The tools of tomorrow depend, not on what the people of today want, but on what the tools of today can be used to build. Wants follow possibilities. The technology, this view invites us to believe, is pulling along behind it the great historic institutions of family, church, academe, and state. And third, the farther the process goes, the more complex and powerful the tools, the faster new tasks become possible. And as new capabilities follow fast and follow faster on each others' heels, the harder it is for these ancient valuing institutions to perform their proper function of passing moral judgment, of legitimatizing or condemning. If these institutions are immobilized or subverted by the technology, what replaces them? If the institutions to which automatically we assign responsibility for the direction of our affairs and the ultimate ordering of our worlds have been left behind, what has taken their place? Where do the effective controls of industrial energy lie? Who owns the shovel?

Chapter IV

THE WISDOM OF THE TOOLS

And the people bowed and prayed
To the neon god they made…
Paul Simon

In his extraordinary book, *The Technological Society*, Jacques Ellul suggests that only two economic pathways lie before highly industrialized peoples. He identifies these as the planned economy and the corporate economy. He discards as incompatible with the irrevocable compulsions of "technique" a third alternative, liberal interventionism. In the first, the planning is done by the state, in the second by great corporate firms. Ellul goes on to demonstrate that the end result is the same, since the commitment in each case is *a priori* to technique. Technique contains its own imperatives.[34]

Galbraith and Gerschenkron have reached the same conclusion, with planning the key concept. They agree that planning is by the technicians, not by anyone else. Plans of the commissariat, in the case of the Soviet Union, or of the President's economic advisors in the case of the United States, either conform with the requirements of the organized technology or are unavailing.[35] If planning which originates outside the technology fails to serve the ends of the

---

[34] Translated from the French by John Wilkinson. Alfred Knopf, New York, 1967. First published as *La Technique ou l'enjeu du siècle,* by Librairie Armand Colin 1954, pp. 183-8.

[35] *The New Industrial State*, Houghton Mifflin, Boston, 1967. Signet edition, 1968. The point which is made in Galbraith's second chapter, "The imperatives of technology," Gerschenkron found to apply to the Soviet Union with equal force, and identical rationale. The technical tasks, once undertaken, make non-technical considerations extraneous. Gerschenkron, "Industrial Enterprise in the Soviet Union." In Mason (ed.), *The Corporation in Modern Society*. Atheneum, New York, 1966.

technology, then the ends of the planners are compromised also and hasty steps will be taken to bring the plans into line with the technological imperatives. If planning is inevitable to technology; if the planning must be the planning of the technologists; then technology is autonomous. This is precisely the logic of the tools-to-make-tools property of the industrial phenomenon which can be extracted from the perceptions of Buckminster Fuller. It is also the conclusion of Ellul. He holds that it is not only autonomous but automatic in its rates and direction of change:

> A surgical operation that was formerly not feasible but can now be performed is not an object of choice. It simply is. Here we see the prime aspect of technical automatism. Technique itself, *ipso facto*, and without indulgence or possible discussion, selects among the means to be employed. The human being is no longer in any sense the agent of choice.[36]

And
> Technological activity automatically eliminates every non-technical activity or transforms it into technical activity. This does not mean, however, that there is any conscious effort or directive will.[37]

I find great excitement in the convergence of ideas of these three extraordinary men: Fuller, a high priest of technology; Galbraith, an agricultural economist assaulting the cliches of industrial definition; and Ellul, a French professor of law, philosopher and theologian. They agree that technology in its present stage creates its own imperatives and goes its own way, regardless of the forms of the traditional normative institutions. They agree that its energy, both mechanically and in its organization, is

---

[36] Ellul, *op. cit.* p. 80.
[37] *Ibid.*, p. 83.

automatic and self-augmenting. They agree that it matters little the names given to the form of organization, whether corporation or commissariat, since the requirements of the technologists derive from the same matrix of tools, with the same capabilities for doing the same kinds of tasks.

If we add to these areas of agreement one which Fuller's mechanical contentions imply, that of technology as a single web of tools, growing on its own edges, the result is industrial society, invented by Western European Man. While it has shaped the human condition with savage violence; while thousands of cultural groups have their languages consigned to the historical ash heap along with their gods and identities; while it has brought the human species bumbling to the brink of its own destruction; its unique properties have gone unnoticed. It may be that trying to describe the industrial phenomenon in terms of its own realities is an exercise in futility. So Ellul seems to think. He concludes that nothing, short of the total collapse envisioned by Brown, can interfere with technology's own headlong course.[38] He holds that presently proposed interferences are hopelessly naïve, their degree of optimism being the precise measure of their nonsense. "Anything and everything which technique is able to produce *is* produced and accepted by the consumer. The belief that the human producer is still master of production is a dangerous illusion." [39] Again, "It is vanity to believe [the monolithic technical world] can be checked or guided." [40]

The imperative of technique, the logic of tools-to-make-tools, is clear. What to do with it is not. Ellul sums up the predicament so starkly that he leaves scarce room for wriggling:

---

[38] *Ibid.*, p. 89. "… there is never any question of an arrest of the process, and even less of a backward movement. Arrest and retreat only occur when an entire society collapses."
[39] *Ibid.,* p. 93.
[40] *Ibid.,* p. 428.

The autonomy of technique must be examined in different perspectives on the basis of the different spheres in relation to which it has this characteristic. First, technique is autonomous with respect of economics and politics. We have already seen that, at the present, neither economic nor political evolution conditions technical progress. Its progress is likewise independent of the social situation. The converse is actually the case...Technique elicits and conditions social, political and economic change. It is prime mover of all the rest, in spite of any appearance to the contrary and in spite of human pride, which pretends that man's philosophical theories are still determining technical evolution. External necessities no longer determine technique. Technique's own internal necessities are determinative. Technique has become a reality in itself, self-sufficient, with its special laws and its own determinations. [41]

And what is the role of men in this one vast integrated circuit?
Ellul again:

We have already seen, in connection with technical self-augmentation, that technique pursues its own course more and more independently of man. This means that man participates less and less actively in technical creation, which, by the automatic combination of prior elements, becomes a kind of fate. Man is reduced to the level of a catalyst. Better still, he resembles a slug inserted into a slot machine; he starts the operation without participating in it. [42]

The industrial control system must not be inhibited by values contrary to those that would rationalize the free

---

[41] *Ibid.*, pp. 133-4.
[42] *Ibid.*, p. 235.

elaboration of tools. If unity and order are the soil of the technological world, an ethic derived from its technical requirements is its climate. What is possible is desirable, necessary, even inevitable. This is the wisdom of the tools, demanding acquiescence in advance to the desirability of whatever work is made available by the next generation of tools. Illustrations abound every day. Picking up tonight's paper in Eugene, Oregon [43] to see what it would yield along this line, I find that a group of newsmen and power company executives (and through the paper, the public) heard the following from faculty members in the Oregon State University Radiation Center in Corvallis, Oregon:

...Chih H. Wang, director of the center, after remarking, "I, for one, would not like to be colored as a friend of industry," went on to cite "the so-called information gap" that exists on nuclear power subjects.

"This is particularly serious," he said, "if one realizes that the public as a whole is inclined to display fear against technological advancements.

"The situation is even worsened when opportunists, making use of the information gap, write sensational books to influence public opinion."

Wang's comments were later elaborated upon by various power company executives, who contended that the critics of nuclear power use authoritative quotations out of context and who implied such individuals are motivated by the desire for notoriety or a quick profit.

Lamar P. Bupp, professor of nuclear engineering, spoke on generation of power by nuclear means and dwelt at length on safety devices incorporated into the design of nuclear power plants. Arthur G. Johnson, radiation health physicist, detailed steps that are taken to minimize radiation hazards and of other sources of radiation with which the public comes into contact.

---

[43] *Register Guard,* July 19, 1968, p. 1A.

"The United States," Bupp said, "is irrevocably committed to nuclear power." The chance of a nuclear explosion occurring from some malfunction of a nuclear power plant, he said, "is categorically impossible."

Bupp also derided other allegations of future accidents at nuclear plants, stating at one point, "The requirements on this are almost ridiculous on the side of safety."

The one-day seminar was jointly sponsored by the Eugene Water and Electricity Board, Portland General Electric Co., and Pacific Power and Light.

As a public relations plug for the ethic of technology, this would be hard to improve upon. It has everything. Beginning with the implication that lack of technical information is responsible for public fears, Wang proceeds to the curious notion that "the public as a whole is inclined to display fear against technological advancements." Conventional rationality must label this sheer nonsense. The *American* public inclined to be fearful of technology? Good heavens! Again,"...technological advancements." The technologically new is by definition an advancement, whatever it is or whatever it means. The professors and the corporate executives were in agreement that any attack on technology (none is cited) is irresponsible and ill informed. Again, what is technologically desired is good. Any voice to the contrary speaks from anti-social motives. By definition. Not just in this instance, but always, in principle. (So much for the difference between science and technology, and the traditional questioning attitude of either.) Professor Bupp's assertion that "the United States is irrevocably committed to nuclear power" can hardly be disputed out of context, given the present investment in such installations; but in the context of this seminar, sponsored by three utility firms, it becomes something else altogether. It caries the implied addendum of "when, where and in the amounts that the specialists deem desirable." Again, a bald declaration of the preeminence of technologists, enunciating the public interest in the words of the managers of the technology. Concern for the human organism, in all its corporal and psychic vulnerability, is by definition being served by the wisdom of the tools, and any claim to the contrary must be uninformed because it is not rooted in that one true wisdom.

"Steps taken to minimize radiation hazard" must be adequate or the technology wouldn't proceed, is the implicit message. But who knows? Certainly not the technologist. Such "sensationalists" as Commoner and Carson have disposed adequately of that fiction. Taken as a whole, this is a set piece of propaganda, directed by a single-minded coalition (or brotherhood) of scholars and corporate managers. The target is the public, through the media representatives. The message is the total wisdom of the tools, and the malfeasance of those who challenge that wisdom.

Surely this could not be mistaken for an isolated example. "Better Living Through Chemistry"; "Progress is Our Most Important Product"; "Peace is Our Business" (over the gateway to Malstrom Air Force Base); "All-State is All You Need to Know About Insurance;" the official rhetoric rationalizing the Apollo Project; the message is the same.

The technological milieu, then, is the one in which the wisdom of the tools *must* prevail. Otherwise constraints on the natural growth on the edges of the tool system would arise from all kinds of subjective sources, having nothing to do with the imperatives of increased busyness. Long range predictions, the orderly introduction of new tools to the art, the orderly development of markets for new products to finance amortization of existing tools and their elaboration; all are jeopardized by the intrusion of any wisdom but that of the tools.

No matter what the particular type of busyness, all industrial tooling shares the necessity for this milieu. Commissariats and industrial communes share it, or will whenever they undertake technology of sufficient complexity. Professor Wang's plaintive protest that the

42

American public tends to fear technology is a bitter irony in light of technology's preeminence. A public that remains quiescent while its mother's milk becomes definable as unfit for human consumption; while its national bird, along with numerous other species, faces sudden and unpredictable extinction from pesticide poisoning; while its greatest water bodies turn anaerobic; while its great cities are bathed in unbreathable air; such a society is hardly suffering from excessive technological nervousness, the current turbulence among its young notwithstanding.

The climate or milieu necessary for robust technology is perhaps primarily one of language usage, and derivative public habits of thought. A state of pollution of the public language seems called for that contains the proper mix and consistency so that the pronouncements of Professors Wang and Bupp and their corporate colleagues will show up in dominant public opinion as information; so that enough of the public accepts the wisdom of the tools that the rest doesn't matter.

Chapter III was an attempt to suggest the shape of the industrial tool system by projecting the logic of its special mechanics, as defined by Fuller. Fuller is a machine-oriented man. When he uses the term "tool," he implies hardware. The notion need not be thus restricted. Procedures may be tools, as in double entry bookkeeping or computer programming. This more open concept is implied by Ellul in his term "technique." Robert Merton, in his forward to Ellul's book, says "by technique...he means far more than machine technology. Technique refers to any complex of standardized means for attaining a predetermined result." This may serve as a general guide to the use of both terms here.

Fuller's view of the mechanical frame of industrialization is powerfully reinforced by those who have approached the problem via the social institutional route. Galbraith, in studying the American corporate system; Gerschenkron, examining Soviet history for western

43

likeness; and Ellul, from a French base studying the entire ramification of applied technique; conflict in no significant way with the projection from Fuller.

Fuller's conception of tools-to-make-tools and self-augmentation of capability provide the mechanical rationale for the necessity for planning and freedom of action found to exist in the technology by the others. What Ellul calls "the autonomy of technique" and Galbraith "the autonomy of the technostructure" share their validity with the Fullerian notion of the nature of industrial tools.

It is time for a sketch of what lies ahead. The work of these students and others carries us to a certain point, given its ultimate credence by non-potability of mothers' milk. This is the specter of a world out of control; a world in which an irresistible, self-directed technology is rushing blindly toward catastrophe; in which human intelligence has been subverted to the service of the acceleration of the tool system, and rendered incapable of exerting subjective influence on direction; in which nuclear bombs and anthrax are officially held to make life more secure; and routine environmental poisoning to make it more rich. Those who seek to understand it have achieved much, to be able to describe it thus far. In the end, however, they stop short. If species survival is the minimal rational goal for studying the human estate, as I hold it to be, and if species survival is placed in imminent jeopardy by our own hands, then what in fact are the human controls? I refuse to accept the notion that there are none, if only because it is intolerable. It may be that my friend is right who argues that human intelligence simply fails to increase as fast as the crises; but I am looking for a rationale for action. We may be butterflies caught on technology's pin, but let us wave our wings a bit longer.

When he concludes that we must individually combat technique, Ellul is advising, then, personal therapy; and it is good advice. Smatterings of youngsters all over the world are following it. They are getting gassed, clubbed and killed for it. But they are still following it. The youngsters are also waiting for the specialists to get on with their work so that their lumps might mean something. But this is hardly enough. Ellul hasn't gotten to the heart of the problem, which is, how does the thing work? Why is public opinion insane? By what mechanism has it been made insane? If public opinion can be made mad, can it not be made sane? But how? Men direct technique, and love their children. At the same time. Let us cling to that. From Fuller, Ellul and Galbraith, it may be possible to go a bit farther. It may be possible to discover ways to subvert the technology to its own salvation.

Men do act. And part of the tool system is made up of social organization. If Ellul is correct, the operational social organization is not what it seems. It is not politics and government of conventional talk. For ten years I taught the mythology of that government to bemused children, who found out sooner than I that they were bemused. There is no alternative to finding out what the realities are. By the same token that they feel the horror, the children have been educated by the technology, given their literacy by its lights and predicates. Michigan Avenue, summer 1968, and Haight-Ashbury, can do little more than provide more opportunities to elaborate the technology of protecting the freedom of action of technology. The heart of the matter must be understood, and it must be in terms ultimately of men. Fuller, Ellul and the others are close enough to the hart of the matter, however, that one must go through them, not past them, to go on. If Congress is not the scene of action, then what is?

Only one institution, in the heartland of the industrial revolution, can claim *ownership* of the tools. Only the same institution can claim the crystal clear freedom from sentiment necessary to manage a system of applied power which can rationalize its own total destruction, and name it progress. That institution is the corporation, in all the elegance of Galbraith's portrait. And it is the corporation, viewed from outside its own public relations cliches and the cliches of impotent government (which after all are the same words), where the heart of the matter most reasonably is to be sought. On with naïve wriggling.

In the west, the elusive corporate institution has emerged relating in some fashion to nation states. Let us examine it here, where it is most accessible, and wherein much spadework has been done. The corporations of concern are not the small firms, making up a corporate population of a quarter million or so in the United States alone. Rather they are the great technical empires, perhaps *Fortune*'s 500 domestic and 200 foreign; perhaps those whose model Drucker was presuming to study by studying General Motors;[44] perhaps Galbraith's "mature" as opposed to "entrepreneurial" corporations.[45]

Assuming the views of Fuller, Ellul and Galbraith to mirror reality; and assuming that industrial technology is essentially in the hands of a few hundred great corporations; then any attempt to fit these entities into the ordinarily accepted social-political-economic scheme of things produces instant chaos. The language simply doesn't fit. If technology is autonomous, then its managing institutions are autonomous, subject in the crunch to no imperatives but those of the tools. Corporations, therefore, must be escaping constraint by national domestic law. Presumed to be subject to regulation and supervision by the state, such regulation and supervision must be sham. Presumed to draw existence from the state, they must transcend the sovereignty of the state and hold sovereignty or some equivalent independently. Presumed to be nationally identifiable, as American or French corporations, such identifications must be illusion. If technology, administered by corporations, is autonomous, then one should be able to show that corporations create their own world of operation, with its own internal logic and imperatives. Perhaps there is a corporate ecology, the

---

[44] Peter Drucker, *The Concept of the Corporation.*, John Day Co., 1946.
[45] *The New Industrial State, op. cit.,* p, 103.

46

examination of which might cast light upon the corporate-technological carelessness for the organic ecology of the natural world which places our species in the same kind of jeopardy which has overtaken our feathered and crusty companions of the poisoned ocean margins.

Galbraith's conclusions in *The New Industrial State* include the notion that ultimate decision-making power lies in the assemblies of specialist-experts deep within the corporate structure. Not the stockholders, whose relations to the great firms have become so remote and ceremonial as to be managerially meaningless; not the directors, to whom information comes pre-digested as it were, and who are therefore limited to alternatives presented to them neatly packaged from below; but the technicians, from management down, functioning never as individuals but as parts of teams bringing together many technical specialties; these and only these have ultimate power in the mature form of industrial organization. Gerschenkron shows that this holds just as true for the Soviet Commissariat as Galbraith holds it to be true for the great American corporations. The key to this is the notion that all significant decisions require the special knowledge and talents of many men. The more complex the tasks undertaken, the more kinds of specialists are required. Thus power lies with the specialists, not with the nominally public or private persons at the top of the structure who must accept ceremonial responsibility for decisions.

Ellul's conclusions arrive at the same point, by a different road. Though he hardly discusses the corporation, through the nature of "technique" itself, he shows in all circumstances the technicians presenting to the authorities (holders of responsibility) information which renders all but one solution unreasonable; information which the authorities will ignore at their peril.

The necessary logical consequence of this is the rejection of any moral judgment that is not generated out of the technical decision itself. Values which in any way conflict with the conclusions of the existing teams of specialist-experts are clearly unreasonable, irrational, and constitute therefore a threat to the orderly progress of the total system. Implementation of such outside values places all levels of the corporate community in intolerable jeopardy. The reactions of corporations to outside views of the propriety of their actions provide nearly conclusive evidence of the value placed on freedom of action.

When I was in high school in rural Idaho, football was second only to the price of wheat as a topic of community concern. Compared to today, however, little money was spent on it. Any adaptable open space was utilized as a field. A neighboring school for several years played on a field that was equipped, in addition to the goal posts, with a telephone pole somewhere near the forty-yard line. That pole was an obstacle to planning somewhat analogous to the presence of Ralph Nader in the corporate world of General Motors. There is no reason why football could not have developed from the first with such an obstacle in the middle of the field, but it didn't. In the same fashion, autos might have been designed from their earlier day equipped with smog control devices. But they were not. In each case, drastic measures were in order to remove the obstacle to orderly conduct of the established system. In each case, changes not originating from within the system were by definition unreasonable and to be resisted on principle.

Ralph Nader was a threat not only to the orderly marketing of the Corvair and the design of exhaust systems. He was a threat to the precedent that innovations in automotive engineering *must* originate from within the in-dustry, and be subject only to the judgment of the industry. [46] The principle is freedom of action. The ultimate corporate

---

[46]Ralph Nader, *Unsafe at Any Speed,* Pocket Books, New York, 1966.

dogma (which is, in Ellul's context, the dogma of "technique" in whatever institutional setting) holds, then, that wisdom can only arise from the internal processes of corporate decision-making.

A tool complex that can grow only on its own edges, in which "...the preceding technical situation alone is determinative,"[47] makes such a technical-corporate ethic inevitable. It must provide a rationale and legitimacy for whatever is necessary to permit orderly planning for amortization of the existing tool complex, while at the same time building the next generation of tools, whatever those turn out to be. Since the possibilities are never wholly predictable, the future must be kept open for the utilization of the new tools, whatever they turn out to be. Corporations must defend their freedom to use the tools, but also to create a demand for their output, *whatever that may turn out to be.*

The military budget for next year is almost identical with the amount of capital scheduled for investment by the private sector. Each is in the vicinity of $80 billion. Both amounts are extracted essentially as compulsory savings on the part of consumers, to be used by institutions. If one abandons the fiction that great corporations do not govern, both amounts may be viewed as taxes. In the case of the military budget, the taxes are visible and direct. Corporate taxation is invisible and masked behind a plethora of names. It is simply added to the price of goods and services, in an administered market. A competition is thus established between sovereign agencies and the individual, to bring all the individual's paycheck under institutional control. In theory, the governmental tax is intended to serve some subjective set of purposes called "the public interest..." The corporate tax is intended to finance the orderly elaboration of the industrial tool system. This is in addition to "profit," a category of industrial earnings that is of declining

---

[47] Ellul, *op. cit.,* p. 90.

significance.[48] Since the previous tools govern what possibilities exist for elaboration at any given time, the corporate interest can be described as that of escalating their own ongoing kinds of busyness. This requires the expenditure of energy and talent on the task of keeping their futures unfettered—that is, to guarantee public acceptance for whatever they will find themselves doing as the technology dictates change.

The governmental tax money is, however, part of the loose money as viewed from the vantage point of the corporation. That part of the governmental budget not spent for salaries is used for the procurement of goods and

---

[48] To the extent that earnings are used for capital expansion, the term "profit" is robbed of its meaning, as the concept is altered in its financial role. The word appears in corporate rhetoric to mean "that category of earnings which is reported to the stockholders as available for dispersal as dividends." The *1969 Annual Meeting of RCA Shareholders* contains the following:

[In 1968] after-tax earnings were $154 million. . . .

We expect to spend more than $200 million during 1969 for new and enlarged facilities and equipment. . . .

Shareholders Mrs. Leon Brucker and Dr. Maurice Golden recommended increasing the cash dividend. The President said that the present cash dividend is a dollar a share, that the rate has tripled in the last six years, and that a great deal of money is needed to keep developing new activities such as computers. . . [The President] pointed out that it does not make sense to increase cash dividends to the point where borrowing at high rates is necessary for working requirements. . . .

Thus "working requirements" come to include "developing new activities"; and the notion of profit drifts away into the semantic fog.

Nor is it necessary to assume that "tax" as a component of price is limited to the invested retained earnings of a single company. The "price" of insurance, or of credit provided by banks, includes an amount that might be considered "tax" to provide money for the capital market. The general lesson to be learned from this is of supreme importance. If the language derives from a theory which is specious (the theory of perfect competition and customer sovereignty), but persists in use; then the function of that language is to conceal what is going on, rather than explain it.

services. Debt service may be included in the general notion of procurement, since interest on public debt may reasonably be viewed as the installment purchase of money. The government's customer function, then, to the corporation resembles that of any other consumer. The corporate community must simply include government agencies and personnel among its propaganda targets, and convince government that it needs what the current generation of tools is producing, and that it will in the future need precisely what the next generation of tools will be designed to produce, whatever that turns out to be. The corporate owners of the tools have, of course, no interest *per se* in what they will in the future be producing, only in amortizing the present generation of tools and preparing the way for the next.

The net result of this is something like the following: when I purchase a package of cigarettes, I pay for a host of things. I pay for the cost of manufacture and distribution, including varying standards of living for all those involved. I pay for a continued national propaganda campaign directed against myself to make certain that I will continue to purchase more cigarettes so that the present enormous complex of tools, including the propaganda tools, which surround the manufacture and marketing of cigarettes, may be amortized in an orderly fashion. I am paying for a separate propaganda campaign designed to so subvert my political system that no concept of "the public interest" which has its origin outside the industry will find expression in public policy. I am also paying for an accumulation of capital that will allow the elaboration of the next generation of tools, whatever those turn out to be. I am also paying for an accumulation of capital that can finance the evolution of a new set of communication tools that will augment the effectiveness of the propaganda. I am also paying for a continued and generalized propaganda campaign to make sure that I do not see, or at least do not get upset by, all that I am paying for. And last but not least, I am paying for propaganda designed to convince me that I am not being subjected to propaganda. All this ultimately serves to guarantee corporate freedom of action; to give free play to the wisdom of the tools. And all this seems to be the doing of corporations. Let us proceed upon that premise, and see if their autonomy can match that of the tools they own.

The only work of consequence derives from fossil fuels, nuclear generation and falling water. In some fashion and at some stage this work is in the hands of great corporations. These creatures of the law of industrial man have come to permeate all aspects of living. Yet they have not been the subjects of searching study, or of a questioning public scrutiny. That their names are household words while they remain shrouded in effective anonymity has not generated a countervailing curiosity. This is most strange. That about 200 of them control half of all the marketed busyness of a society of 200 million people is no secret, but stimulates little concern as to their nature, character or ancestry. They generate, select and dispense what passes for goods, news, books, weapons, scenery, propaganda, landscape, vice and virtue.

They hold the airways in usufruct ownership, and the means of energy transmission in fee simple. They have become indispensable, omnipotent and as invisible as the sky. They have names but not substance. Virtually all wealth passes through their hands, and virtually all goods and services are ultimately purchased from them. They are not taxable or constrained by public law. The policy they make in private secrecy surfaces, masquerading as public policy.

During the years of corporate adolescence, when they were unsure of their powers and clumsy in their actions, lack of discipline was tolerable. The broken social china was offset by their unique capabilities to transcend the old conservatisms, which got in the way of elaboration of the new machines. But like Topsy they growed, in size and sophistication. They learned to hide in full view, and that the more of the landscape they filled the harder they were to see.

They learned to escape man's insanities, such as nationalism, and his sanities, such as concern for breathable air. They learned that the public language can be restructured from a vehicle for information into a vehicle for non-information, and that control of the public language is the surest road to corporate freedom, which is to say health, wealth, privacy and power.

The energy pouring through the airways, pipelines, factories and publications of the great corporations today is so great that it defies, indeed overwhelms territorial limitation. All of mankind can be addressed simultaneously, and simultaneously poisoned. Whoever objects can be hired or safely ignored. Nothing of significance escapes, at some stage or level, the hand of corporate power.

None of this is intended as criticism of the corporation in the sense of moral obloquy. The corporation is the creation of sentient human beings, as are all institutions. If the corporation is used to organize and administer the energy that terminates the human experiment, it is man, or rather men, in the persons of European-Americans, who will have led their genetic brethren along the same path down which they have sent so many other species. Corporations are not our enemies. Ignorance of our own tools is our enemy, and curiosity our only weapon. No student of the predicament of man can ignore with impunity the controls over industrial energy. And until the ignorance abates, it belongs equally to all.

Theory of government that fails to include the institutions controlling the work capability of the world is defective theory. A geography incapable of dealing in systematic fashion with the institutions which determine and implement the rate and quality of landscape change is hardly a complete geography. An economics that ignores the great agents of production, distribution and propaganda lacks explanatory power. Charles Olson, the Gloucester poet, wrote, "...I have had to learn the simplest things last/ Which has made for difficulties..." The difficulties we face from not knowing our most obvious institutions are likely to be terminal. This ignorance has permitted deadly courses of action to become deeply imbedded in the fabric of our societies. That species survival is a significant notion seems simple, but we are learning it last, if we learn it at all.

Chapter V

THE MATURE CORPORATION: PERSON OR SWARM?

Behind each barricade
we man our stations,
Resolute
in the least of things.
Jo Merrill

"Few subjects of earnest inquiry have been more unproductive than study of the modern large corporation. The reasons are clear. A vivid image of what *should* exist acts as a surrogate for reality. Pursuit of the image then prevents pursuit of reality." Thus Galbraith begins his *New Industrial State.* Another reason for frustration is the simple difficulty of dealing with the corporation in the language of other social-political institutions, among which it is anomalous. This task is reminiscent of trying to eat raw egg white with a knife. When the law decreed the corporation to be a legal person, it became in law what it could never be in substance. The people have by statute and judicial inter-pretation created a metaphor and made it to act as though it were animated material. This idea, straight from the other side of the Looking Glass, must be accounted as one of the most powerful industrial tools ever devised. The following attempt to put its operational characteristics in perspective requires patience and forbearance from the reader, since the task is best suited to the talents of a Lewis Carroll, who I am not.

Once the legal metaphor of corporate personhood is an established convention, the tendency to extend its language to include bodily properties is well-nigh guaranteed. Thus if one is speaking in terms of the "person," it is only reasonable to assume the metaphor to be complete; that is, to speak of all properties of the corporation in language of the "person" metaphor. The corporation would be regarded as having not only the political but biological attributes of a person. To deny the latter seems to deny the former; but the former is required by law. Let us then split the metaphor apart, and examine just what the law can convey.

First let us concede the power of the law to define and enforce. The corporation *is* a person, in all ways that lie within the competence of the law to determine. Let us not assume, however, that the metaphor must then be turned back upon the law, and the law be assigned powers of procreation in a biological sense. The law has *not* created a flesh-and-blood person, and the issue is only confused by assuming that it has. Let us then strip away all pretense of biological attributes and examine the corporate self in its resulting nudity.

The most obvious property of the corporate person is the absence of a body. It has, therefore, no forbears, no heirs, and is immortal. It has neither youth nor adulthood, only chronological age and size. It is spared all corporal (as opposed to corporate) fears of injury and death. It may be robbed but not raped, confiscated but not killed. Its lacks sense. Without eyes, it knows neither beauty nor ugliness, nor can it read. Without ears, it knows neither sound nor silence. Lacking genitalia, it is immune to the joys and blandishments of sex. Lacking lungs and liver, it cannot strangle on its own effluents.

Having no body, the corporation escapes all pleasures, vicissitudes and capabilities that depend for their existence on the presence of flesh and blood. Strictly speaking, a corporation can "act" only in a narrowly restricted sense. It may "own," "arrange for" (legally) or "conspire" (illegally). It does not build, it arranges for building to be done by its flesh-and-blood agents. It probably should be considered able to buy, sell and sue. It cannot commit crimes of violence, and imprisonment is not among its spectrum of hazards. It can be guilty of crimes, however, and a variety of criminal acts may be committed in its name. The majority of acts of criminality fixed upon corporations by the judiciary go under the name of "conspiracy." It could hardly be otherwise.

The corporation is, by virtue of its inorganic corpus, singularly free to follow a closed and stark rationality, in contrast to its human counterparts. The flesh-and-blood person who undertakes to live his entire life relentlessly in accord with a closed rationality is, ordinarily, either killed or institutionalized as soon as his aberration becomes apparent. For the natural individual, any tendencies in this direction must be tempered by frequent flights of irrational compassion and alms-giving, and/or the irrational ceremonial equivalents.[49] Even those of immense power, the heroic capitalists of our immediate past, were eventually turned on by society (which, prudently awaiting their death or advanced senility, labels them in retrospect "robber barons"). These powerful adherents of the Newtonian rationality of the "dismal science" of Ricardo and Malthus find their way into the folklore as mighty villains, because the rationality of their technical pursuits came in the public eye to be affixed to their persons. The unembodied mature corporation can pursue such a single minded rationality secure in the knowledge that its natural officers may, in ways of no moment to the corporate purpose, mask the rationality of the corporation with their private participation in the required irrationalities. It is this process which is known as being "a

---

[49] Elllul makes this point in an interesting fashion, speaking of the general notion of technique rather than of technique as the heart of the idea of the corporation. The application seems clearly interchangeable. "In technique, whatever its aspect or the domain in which it is applied, a rational process is present which tends to bring mechanics to bear on all that is spontaneous or irrational." (pp.78-79). And again, "Human life as a whole is not inundated by technique. It has room for activities that are not rationally or systematically ordered. But the collision between spontaneous activities and technique is catastrophic for the spontaneous activities." *op. cit.,* pp. 82-83.

good corporate citizen." A vice-president joins the Rotary Club and chairs the United Appeal fund-raising drive. He is being the corporate vicar to natural humanity, and his action is commonly but wholly erroneously interpreted as reflecting a corporate value. It hardly ranks as an expose to reflect on the serried ranks of public relations men and tax lawyers lurking behind more direct corporate philanthropy. Since the vanishing of the great closely held "entrepreneurial" firm, no man embodies the corporate identity. This not only permits un-deviating rationality. It also releases the full power potential of the concept of the legal person. The secret is one of identity.

A natural man with a natural body may change his name but not his identity. He is himself, the child of his parents, the father of his children, citizen, alien, this or that—but nevertheless one person. A sufficiently diligent effort to discover his true identity commonly serves to do so. Even if it doesn't, the identity is recognized as simply unknown; and, if occasion demands, the person will be buried in a John Doe grave or arrested with a John Doe warrant. John Doe, however, could never be stretched to subsume the intricacies of corporate identity. Above all else, its fluid identity has made the corporation the ideal institution for managing the industrial tooling of the contemporary western world. The corporation is simultaneously by legal compulsion both free person and owned. This is the fact from which corporate power ultimately flows. Let us now poach the egg white in a broth of only partial whimsy, and see if it will cling to the fork.

A corporation cannot exist un-owned, and is a legal person, sharing the general prerogatives of citizenship. This gives it three separate but simultaneous identities. It is as though a pre-Emancipation, free black American owned himself as a slave, while simultaneously being owned by another. (The corporation in effect owns itself, while being owned by stockholders.) If such an individual held the corporate prerogative of presenting himself before the law as freeman, his own slave, or the slave of another as his momentary interests dictated, he would hardly have needed Lincoln as his benefactor. But this is only the basic element. Ramified far enough, it becomes a multiplex giant such as ITT or General Motors.

In its most halcyon days the ante-bellum South did not know a circumstance when a slave could hold title to another slave. But a corporation, owned by its stockholders, can in turn own another corporation, which shares the same three identities as the initial one. Terminology follows the metaphor and labels the first the parent corporation, but succumbs to greater frankness in the case of the second and labels it a subsidiary. Recognition that the identities of parent and subsidiary can be reversed by a simple stock transfer provides a hedge against any notion that there is now a family, a favorite corporate homily. But the term "subsidiary" is far from precise. It is a statement of ownership, but little else. A host of variations on the subsidiary theme are possible, including most prominently the ubiquitous philanthropic foundation; the non-profit corporation; and a multiplicity of varied arrangements that divide functions between horizontal sets of subsidiaries and vertical administrative divisions. The secrecy inherent in the well-nigh universal practice by parent corporations of reporting only aggregate accounting figures for public consumption, allows for effective concealment of specifics of such relationships. The plethora of possible identities would boggle the mind of the most diligent genealogist whose first tool would have to be a sheaf of subpoenas. A conglomerate giant like International Telephone and Telegraph, revealed in FCC hearings to operate through 433 separate boards of directors,[50] is, when it reports to its stockholders, one person. The law, however, must grope

---

[50] Nicholas Johnson, "The Media Barons and the Public Interest," *Atlantic*, (Vintage edition, 1969, p. 44.

through an amorphous swarm to fix responsibility for a transgression.

Another general dimension should be added to the problem of identity, that of nationality. Vernacular and formal language, as well as the law, assign nationality on the basis of the state from which the corporation's charter derives.[51] *Fortune*'s division of the world's corporate elite into American and foreign lists is taken quite for granted and seems to have no reasonable alternative. But neither does such division stand close scrutiny. The corporation which has charters drawn from a single state apparently is, in its personhood, a citizen of that state. But who owns its stock? It may be that a corporation is a citizen owned by citizens of other states. Indeed, there is the logical problem of property being citizen at all. It might be argued that the cultural backgrounds of corporate managers convey *de facto* if not legal nationality. Standing in opposition to this view is the whole argument of the imperatives of the tools. Corporate culture is the manifestation of this single phenomenon and it matters little whether board meetings are conducted in English, French or German. Style may differ, but not substance. Suspicion persists that corporations may adopt or reject nationality at will, as with other types of identity.

Bearing in mind that each subsidiary exists by virtue of its own charter, the constraints of nationality (though none of the benefits) appear even more illusive in situations of greater complexity. Consider the basic case of a parent firm with its charter deriving from state A, in controlling ownership of the stock of a corporation whose charter derives from state B. This places the second in the status of subsidiary, wholly subject to policy discipline of the first, but leaves its nominal nationality, as a citizen of state B, undisturbed. Thus the parent corporation has access to the prerogatives of citizenship in two states, in each of which it

---

[51] Walton Hamilton, *The Politics of Industry,* Alfred Knopf, 1957, (Vintage edition, 1967), p. 111.

has the option of claiming citizen or alien status, whichever is to its advantage. Any natural person trying prestidigitation like that would find himself sailing endlessly back and forth on the Hong Kong-Macao ferry, or in someone's jail. Yet the realities of firms such as ITT, Ford Motor Company, IBM and scores of others, multiply this complexity by tens or even hundreds of times.

Variations on the subsidiary theme are worth particularly careful examination, if one is to get some kind of sense of the freedom of action this ingenious instrument can provide. Subsidiaries may be created by their parents for special purposes, or acquired through stock purchases. They may serve primarily as internal holding companies within which to group other subsidiaries, or they may be operational, or both. A common practice is to create a holding company into which is grouped all, or all of a type, of foreign subsidiary firms. This allows separate recording of foreign and domestic sales, without revealing actual sources. Examples are IBM's World Trade Corporation, of which all foreign firms controlled by IBM are subsidiaries; and ITT's Worldcom, into which certain of ITT's international communications firms are grouped. Some form or another of this device seems to be standard practice among corporations with extensive operations in more than one state.[52]

Although the full corporate use of these intermediate-level holding companies is scarcely available to an outsider for description, certain potential values are clear, besides the opportunities for bookkeeping secrecy alluded to above. But an entity may serve as a cushion against assaults on company assets, policy or personnel in foreign firms. The holding company can be presented to the publics involved as the parent of the offending firm, thus keeping trouble unconnected with the true parent. It can diffuse the factor of

---

[52] These holding companies are almost always mentioned in the Annual Reports. It precludes the necessity of mentioning individual foreign subsidiaries.

nationality, as viewed from either end of the chain. It can serve as the contracting agency with foreign governments, either carrying out the contract itself or subcontracting to its affiliates in the contracting state or another; but in any case leaving the parent nominally uninvolved. The parent may even be used by the subsidiaries as a subcontractor. In short, it multiplies exponentially the assortment of choices available to the parent in any kind of transaction. Consider the use of the holding company for foreign affiliates in light of the following extract from the 1967 Annual Report of ITT:

"During the year, ITT continued as a major supplier of services and sophisticated equipment for air navigation and communication, including satellites and defense electronics, *while its affiliates abroad contributed to the defense requirements of their own countries*." (Emphasis mine.)

The potential conflicts of interest implicit in such a circumstance are clearly limitless, if one were in a milieu wherein interests might conflict.

The most obvious one involves security. "Defense requirements" certainly must mean involvement with classified material and knowledge. "Their own countries" must mean states from which affiliates draw chartered existence. How can the governments at either end of the chain prevent the sharing of critical technology through as much of the corporate structure as is convenient to the corporation? Will a circumstance be tolerated wherein the admitted technical capability is less for one country than for another? Does it solve the problem if the information and technique is classified by both governments? Precedent in World War II, as described by Walton Hamilton, would indicate that the intermediate holding company might be dissolved (or disappear into the safe), thus creating the illusion of independence of the foreign subsidiary for the duration of hostilities. Hamilton describes just such a situation, except that the parent corporation vanished for the duration, since there was no intervening entity.[53]

Corporations have had somewhat more than a century to work out ways to play this game and still maintain a facade of nationality, since the early experiments of the Krupps in Germany. But Krupp was subject to the same limitations as any great closely held firm. Krupp persona embodied the Krupp firm, and drew onto themselves the public obloquy directed at the firm. Corporations had transcended this crude limitation by World War II, albeit not with anything like present sophistication. The elaboration of layers of subsidiaries has certainly been one of the most rewarding ploys.

Another subsidiary arrangement is well illustrated by Western Electric, the wholly owned manufacturing arm of AT&T. Western Electric's original purpose was to manufacture telephone equipment for what primarily was a communications utility. Today it is much more. It serves many purposes, including that of masking the size and extent of its parent, the world's largest corporate enterprise. AT&T used its subsidiary, Western Electric, as the contracting agency for government work in communications, aerospace and weaponry. Western Electric is one of the few subsidiary corporations which issues annual reports separate from its parent, thereby strengthening the illusion of independence. Research and development for government work is carried out in a group of non-profit subsidiaries whose control is shared equally between AT&T and Western Electric (Bell Telephone Laboratories and Bellcom), or wholly owned by Western Electric (Sandia Corporation). Bell Telephone Laboratories was originally the research center for the service and manufacturing operations of AT&T and Western Electric; while Bellcom and Sandia were formed specifically

---

[53] Walton Hamilton, *Ibid.,* pp, 124-5. Hamilton's Chapter 4 is a mine of information concerning devices by which corporations manage to transcend national boundaries, and turn the restraints of nationality into extraordinary freedoms.

to do government contract work, Bellcom for NASA and Sandia for the Atomic Energy Commission. (We will return in a moment to the curious notion of a non-profit subsidiary of a profit corporation.) Bellcom's principal interest centers at the Cape Kennedy, Houston and Hunstville (Marshall) space flight centers. The Sandia Corporation operates two laboratories for the AEC, one at Albuquerque, N.M and one at Livermore, California. (For some reason which invites speculation, the location of these laboratories was last mentioned in the corporate reports in 1963.)[54]

Nothing more clearly reveals the possibilities of diffusing (or defusing) public awareness of corporate involvement in massive weapons projects than the ABM controversy, which has waxed oftener than it has waned for the 10 months prior to the time of this writing. The Safeguard antiballistic missile program is the current stage of a continuous R&D effort that began in 1945. By the late 1950's it had produced the Nike-Ajax, followed by Nike-Hercules, which were deployed around numerous American cities as a bomber defense; and a more ambitious effort in 1955 to develop an anti-missile missile. From 1945 to 1968, some $4 billion were spent on the project. From its inception, Western Electric has been the prime contractor, with its subsidiary, Bell Telephone Labs, the major R&D agency. In its 1963 Annual Report, Western Electric claims credit for the idea of an anti-missile system for Bell Telephone Labs. From Nike-Ajax through Nike-Hercules, Nike-Zeus, Nike-X, Sentinel and Safeguard, Western Electric (which is to say AT&T) has been the prime contractor. The public is hardly likely to be aware of this, however. Outside of the Annual Reports and the corporate indexes (*Moody*'s, *Standard and Poor*'s), only once have I

---

[54] The above described relationships were gleaned from *Moody's Industrials* index, and the Annual Reports of AT&T and Western Electric, 1963-1968.

seen the name of AT&T linked with ABM in any way.[55] In the book, *Anti-Ballistic Missile: Yes or No*, published by the Center for the Study of Democratic Institutions, reference is made to the first feasibility study contract for Nike-Zeus with Bell Telephone Labs. This is the only clue in the entire book as to who is responsible for the project, and there is nothing to link Bell Telephone Labs with Western Electric and AT&T, whose names appear not at all. Only in the *New York Times Magazine* for May 4, 1969,[56] have I seen clear identification of the contracting firms and their corporate affiliations. Since the omission of the corporate names from the discussion is so universal, not only with the press but with Congressional and administration figures, friends and foes of ABM alike, I can only conclude that the proliferation of subsidiary corporate units and frequent changes of project names has been successfully designed to keep attention focused on the army and the Department of Defense. Any glance which strays toward the private participants can be intercepted by one of the two levels of subsidiaries, leaving the parent wholly uninvolved, as far as the public is aware, in what promises to be one of the bitterest controversies of the decade. As in the international case, the subsidiary can buffer the parent against dangerous and unwanted shocks. The name of this game is an endless pursuit of devices to obscure, diffuse and hide what is really going on, so that an advantageous version of it can be plausibly presented, whatever situation arises.

Let us return now to the topic of the non-profit subsidiary. The approximate difference between a profit and non-profit firm, subsidiary or not, is that the non-profit firm reinvests all instead of part of its earnings above operating expenses. It substitutes growth for dividend distribution. Non-profit firms which do business with the government on negotiated plus fixed fee contracts do not forego the fixed fee (contracted-for profit margin). They use it for

---

[55] Ralph E. Lapp, "A Biography of the ABM," *New York Times Magazine,* May 4, 1969, p. 30.
[56] *Ibid.*

expansion.[57] Along with dividends, of course the non-profit firm may eschew stockholders' annual reports. Since the standard corporate indexes are guides to the publicly held firms and published for the primary readership of current and would-be stockholders, the library researcher who seeks to get a general notion of what goes on in the non-profits faces a formidable task. The current undertakings of Rand, Aerospace, or The Hudson Institute come to light through press releases or otherwise, at the discretion of the firms. The activities of the thousands of independent non-profits, many of them in "think-tank" activities, proceed in stygian darkness. Those that exist as subsidiaries of the great commercial firms may, if they are carrying on spectacular activities such as the Bell Labs, be given considerable space in the Reports. (Typically AT&T devotes one page to

---

[57] H. L. Nieburg, *In the Name of Science*, Quadrangle Books, Chicago, 1966, pp. 254-255.

This is an extraordinary book, detailing the workings of the contract system as it relates government and corporate entities. He shows with great clarity the mechanisms by which the generation of public policy can be transferred to private institutions, there protected by the inherent secrecy of the corporate world. The effect of this is unmistakably in line with Ellul's view of the imperatives of technique. The technological arms of government share the same properties as other parts of technological organization. They must remain free to elaborate new tools, unencumbered by the subjective judgments of legislative action. First the Air Force, then other agencies, have seen the opportunities to transfer parts of their functions to the dark environs of corporations through the device of the negotiated contract. The Air Force most potently, but later others, found that they could generate new non-profit firms for this purpose, where suitable ones did not already exist. Relatively small amounts of money in the early stages could be used to create a snowballing effect, by accelerating some highly specialized aspect of technology and making its possibilities look necessary to the national security.

Galbraith, in his treatise "How to Control the Military" (*Harper's*, June 1969, pp 31-46) says of Nieburg's book, "This is a book of first rate importance which the author was so unwise as to publish some three years before concern for the problems he discusses became general. But perhaps he made it so."

government and defense work, the bulk of which involves their big non-profit affiliates.) But independents or subsidiaries of non-commercial or non-publicly held organizations such as universities are likely to be quite inscrutable.

That there could be a non-profit subsidiary of a profit corporation is a curious phenomenon. Since the General Accounting Office has such a difficult time sorting out the financial arrangements in the corporations doing business with government, a layman can only speculate about the actual tax and other financial benefits that might accrue from including non-profits in the affiliate stable. The relative ease with which money is moved through a corporate hierarchy of firms[58] would indicate that the direct financial advantage of the non-profit to the parent might not be worth the trouble. Public relations benefits derived from the non-profit name, and the enhanced possibilities for secrecy (or perhaps better, corporate privacy), might very well be worth the trouble. The Sandia Corporation protects the image of AT&T and of the University of California from involvement in atomic weapons development. In this case, the non-profit is the subsidiary of a commercial corporation, and provides a link of low public visibility between AT&T, the Atomic Energy Commission, and the University of California. The Stanford Research Institute and similar non-profits serve the same function, but from the other end. SRI is a subsidiary[59] of the University, doing contract work for both the government and private firms, and its work gets reported nowhere in public, except by accident or at its own behest.

The ubiquitous non-profit adjunct has been flushed out of the bushes in recent years by students, Ralph Nader, James

---

[58] Sanford Rose, "The Rewarding Strategies of Multinationalism," *Fortune*, Sept. 15, 1968, p 100 ff.
[59] *San Francisco Chronicle.* April 9, p.1. The status of SRI has been a matter of student interest, and Stanford University President Pitzer is reported to have said that the University owns the Institute. Ridgeway also refers to SRI as a subsidiary of the University.

Ridgeway, and some other gadflies.[60] These exposes have made it perfectly clear that university professors and administrators have acutely tuned in to the Ellulian logic of industrial technology. Whatever research that can be done will be done. Of that research, whatever might embarrass the university if it were made public is likely to be hidden in the non-profit subsidiary research institute, which can maintain an illusion of independence from the university. The relationship between the institute and the faculty can then be regulated to fit the climate and convenience of the moment. Given its corporate status, the faculty may serve the institute as consultants and be paid from institute funds derived from outside contracts, or they may work in the institute in the capacity of university faculty.

In short, the non-profit corporation increases the options of identity of whatever organization to which it is appended as subsidiary. It can give private corporation status to a public university, serving as a sort of institutional strongbox in which to conceal certain aspects of university life. It permits its parent, public or private, to engage in kinds of work and arrangements that might prove embarrassing or impossible to undertake directly.

Since it partakes of the question of identity, the view of the corporation as a self-contained society must be dealt with. I am tempted to believe that, without in any way demeaning the value of studies based on this view, it has seriously impeded the understanding of the larger industrial picture. Perhaps "impeded" is not the right word. Perhaps it simply has pre-empted the attention of the woefully few students that have paid the corporation any serious attention

---

[60] James Ridgeway, *The Closed Corporation.* Ballantine Books, New York, 1968.

Ridgeway's book, though vulnerable to criticism of what he makes of the data, provides an extremely useful outline of use by universities of the non-profit and profit corporate devices.

at all. Drucker's pioneering study of General Motors[61] was principally of this nature, examining the population employed by General Motors almost as an anthropologist might scrutinize the people of a Polynesian island culture. It is within this context that the corporation has been spoken of as a private government. The notion of a private government over employee citizens leads to suggestions that the constitutional protections of individuals be extended to cover the arbitrary acts of corporations against their own in-house "citizens." (This recognizes that the concept of limited government has scarcely entered the corporate boardroom.)[62]

Useful as such exploration is, it commonly fails to take into consideration that the identity of a society made up of a population of natural persons is only one of the myriad identities available to the corporation. A strong case can be made for the possession and exercise by the corporation of something analogous to political sovereignty. That this can be equated with government is not so clear. A state, for example, has the power to tax its citizens. The corporation, sailing in an administered market, has the power to tax its consumers, only a miniscule fraction of which are its "citizens" in the sense of the employees. It collects taxes for the state (not only by withholding; but also to the extent that in an administered market it is passing its own tax along to its consumers). But it also taxes its consumers on its own behalf to the extent that it retains earnings for capital investments. With the elaboration of mass advertising, this taxing power approaches compulsion, not against any specific individual but against the entire target population. This may be a worldwide population, rather than a national or regional one. If, as the flourishing state of Madison Avenue seems to suggest, demand is a manufactured good

---

[61] Peter Drucker, *The Concept of the Corporation*, The John Day Co., New York, 1946.
[62] Representative of this view is Adolph A. Berle's suggestions for constitutionalizing the corporations in "Second Edition/Corporate Power," *Center Magazine,* Vol. II, no. 1, January 1969, pp. 76-84.

like any other, then an administered market seems clearly to offer the opportunity for private compulsory taxation. But the circumstance is foreign to ordinary government, vaguely resembling one in which Belgium, say, was empowered to levy a retail sales tax on some percentage of transactions in the United States; the proceeds of which would go to finance a tax payable to the EEC, and a war against Belgium's neighbors designed to expand her territory. I admit the fit isn't close, but what governmental scenario would fit better? Any attempt to equate the corporation with government is further complicated by the fact that at its will the corporation can slide into its legal identity of one of a number of persons, and cease to have a population.

It would be futile to try to exhaust the possibilities of shifts and combinations of identities made possible by the corporate concept of the legal person. It has been to the apparent interest of industrial society to allow the endless elaboration and sophistication of this device. The early if not original purpose of the corporate form was dual: to allow the agglomeration of capital, and to limit the liability of investors. But in its present stage, with the imperative of maintaining freedom of action in the elaboration of its tools and functions, by far its most important power derives from its ability to be one thing now and another then, one thing to you and another to me, as the situation demands. There is one identity denied the corporation, however. In none of its manifold forms does it have access to the body of a natural flesh-and-blood person, with senses and sentiment, lungs, liver and loves.

It is no accident that I have spoken as though "corporation" means "great corporations." It does. Its references are Galbraith's "mature corporations," the great firms which win places in *Fortune*'s annual lists. The lower limit of size is arbitrary. The upper limit is AT&T, Metropolitan Life, General Motors and the Bank of America, old baronies steeped in respectability. Its *nouveau riche* are Boeing, North American Rockwell, Aerospace, and MIT, living off Department of Defense contracts. Its radicals are the conglomerate upstarts ITT, Litton, Ling-Temco-Vought and CBS.

The reasons for concentrating on these are Galbraith's: they effectively control the bulk of the industrial capabilities of the society, and all of the key elements, such as communication, energy, transportation, sophisticated research and development, metallurgy and finance. Whatever is indispensable to industrial life is to be found in the hands of the corporate elite. As Galbraith, Berle and others have made eminently clear, these share the category of corporations with hundreds of thousands of small firms, but they share few institutional properties. Differences are sharp. Galbraith's analysis will serve as the basis for separating out those members of the total population of corporations that are presumed here to form the critical species. The distinction needs to be made in considerable detail.[63]

The modern corporation came into recognizable form in the early 19th Century as capital overtook land as the critical factor of production. By the end of that century sufficient power had accrued to it that it was a more or less successful competitor with governmental institutions for direction of public policy. At this stage, the owners were the managers, and their names (at least regarding the large firms) were household words. This was the heyday of what Galbraith calls the "entrepreneurial corporations." It largely antedated the marriage between scientific research and industry which was approximately consummated with the advent of World War I.[64] The vernacular, and to some extent, the orthodox picture of what the corporation is derives from this period (Galbraith's "conventional wisdom"). One durable legacy to the modern day is a host of baronial families associated with

---

[63] Galbraith, *The New Industrial State.* The following roughly summarizes Galbraith's corporate portrait.

[64] Fuller, *op. cit.*

71

these firms, who are often propelled by their names and inherited wealth into public office or other public limelight.

These entrepreneurs were masters at concentrating the then critical industrial commodity, capital. As long as capital was in short supply and labor was not, the owners (capitalists) exercised effective control over their firms. Their uniform purpose was the maximization of profits. Therefore the growing interference of government, through taxation or regulation, directly reduced the income to the owner, who, since he was already maximizing profit, could not recoup these costs from the market.

Subsequently, after a short period of labor ascendancy, the research investment triggered profound changes. The resulting institution was what Galbraith refers to as the "mature corporation." This differs radically from the entrepreneurial corporation but has not entirely replaced it. It differs essentially in a migration of decisive power from the stockholder-owners to the body of specialist experts deep within the organization. No man's name is any longer associated with the firm in the public eye. Management and ownership cease to be related except ceremonially. To allow stockholders a policy voice would be disastrous to the functions of the teams of specialist-experts. Most significantly, power is seen to have moved from those who derive income from dividends to those who are salaried. The implication of this is a shift of goal from maximization of profit (which benefits stockholders) to growth of firm, which employs more specialist experts and provides promotion for those already employed.

The entrepreneurial corporation has not disappeared, but has lost primacy in the system. The mature corporation is large, and controls by far the bulk of economic activity of the society. It is the mature corporation that is wedded in its operations to the state, while its officers still on occasion voice the traditional hostility of business toward government which reflected the reality of interests in the heyday of the great entrepreneurs. The real but unspoken hostility of interests today is seen as existing between the relatively new mature corporations in cooperation with government on the one hand, and small business together with remnants of the entrepreneurial corporate world on the other. Both are committed by tradition to polemics hostile to government. Only in the latter case, however, is it spoken from the heart. The mature corporation depends too heavily on government in too many ways to risk serious offense. Government's role in this corporate existence includes the provision, through education, of the necessary pool of specialist experts; and, in part by serving as consumer, guaranteeing against loss in the longer range and more complex technological undertakings. This latter is largely rationalized under the rubric of defense. Non-defense aspects of government protection against loss may in the long run prove to be more profound than massive direct government subsidy through defense research and development and hardware procurement. They include, for example, the public policies relating to the use of television air. These policies have amounted in effect to granting control to the advertising industry (which means to their employers, the large corporations) over what is broadcast; and backing this by providing tax relief for advertising costs.

A central point made persuasively by Galbraith, and which dovetails perfectly with the ideas of Fuller and Ellul, relates to the mature corporation's absolute need for planning. This is occasioned by several convergent circumstances: the necessity to keep in permanent employ functional groups of specialist experts, who are not easily replaced and which may not be substantially reduced in size without virtual destruction of effectiveness; the extraordinarily long lead time from drawing board to market, which makes imperative the effective anticipation of price and demand; and the enormous amounts of capital committed, often years in advance of sales. Thus the mature corporation's first necessity is to find ways to gain control of its own future and pass the cost to the public. These ways have included almost total capture of the mass media, particularly the airways, not only for the purposes of controlling demand through advertising, but also to have at hand a ready mass channel for shaping the political climate; obtaining government subsidy on a long-term basis for research and development (about 90 per cent is government funded); industrial gigantism, allowing administered prices among other kinds of control; and diversification of production, largely through acquisition of firms engaged in a variety of enterprises but also through foreign investment and the consequent development of market areas in diverse political climates. This latter means that unforeseen adversity in one function can be weathered without threat to the whole structure. It also allows additional opportunities for some concealment of activities from public scrutiny and interference.

Hopefully there has appeared a picture in broad outline of two kinds of corporations, distinguishable by organization and purpose, but not in law. The one, representing the older form, is controlled by its owner-stockholders. It includes most of the society's small businesses and by far the greatest number of the nation's some 200,000 existing corporations. Its purpose is the maximization of profits for its owners, bringing it under considerable classical control of the market behavior of consumers. Since it is already reaping the highest possible revenues from the market, any added costs in the form of taxes or government regulation fall directly on the incomes of the owners. Thus the general stance of these entrepreneurial corporations is one of hostility to government, and particularly to any increase of government impingement on their affairs or revenues.

The other is the mature corporation, in which the owner-stockholders are almost totally divorced from management. Since control lies with salaried personnel, the overriding purpose is growth of the firm. Contraction of size is a calamity to be avoided at all costs, since it means breaking up the functioning teams of specialist experts whose decisions are effectively in control and are essentially irreversible at the executive level. Continued growth requires control of future demand and price. Forces marshaled to accomplish this are many and varied. Advertising is relied upon to control demand in the private market, as well as in the creation of a desirable corporate image with the public. Thus advertising serves generalized political purposes as well as direct market purposes. It is used to promote the notion that the market is controlling in the older fashion, and otherwise obscuring from the public what is really going on. Price administration in the private market is served by corporate size and by reduction in the number of firms engaged in production.

The state as consumer provides predictability and reliable planning in as much of the corporate operations as can be diverted to government programs, which for some companies is 100%. More commonly, government contracts account for only a part of the corporate activity. It provides a long-term base, guaranteed against loss, for the organization of teams of specialist-experts and underwrites high-risk technological undertakings. These may or may not contribute privately marketable products, but when they do these can be readily siphoned off, leaving cost of development with the state. The mature corporation also relies on the state and its public education function for a continuous supply of specialist experts. Thus the state and mature corporations are so welded together by their mutual interests and interlocking functions that onerous interference can be prevented, and real hostility is hardly possible. After all, we are dependent on the world's work getting done. And the great corporations do it.

Chapter VI

THE MATURE CORPORATIONS AND THE LAW

...But something is happening here
and you don't know what it is,
do you, Mr. Jones?

Bob Dylan

Society does not dare tamper with the large corporations, except peripherally. This is not because of possible retribution wrought by furious managers. It is simply that the work they are doing is deemed utterly necessary, and there is no alternative mode available in getting it done. This results in a constantly self-perpetuating and reinforcing constraint on constraint. If regulation originating outside the corporate sphere turned out to be excessively damaging, the vital work under control of that segment of industry would suffer; and industrial society is so interlocked in its various elements that malfunctions in one part would tend to feed disruptions through much or all of the system.

The role of the Federal Trade Commission in the regulation of advertising illustrates beautifully the tailoring of constraint to fit the corporate need for freedom.[65] The FTC was created in 1914 in response to the concerns of Congress of that day. Congress was primarily worried about monopoly and ineffectiveness of antitrust legislation. Its mandate to the FTC, however, was rather broad, providing for interdiction of "unfair methods of competition." This was conceived by the Commission to be sufficient to bring falsity in advertising within their purview, and their assumption has prevailed to the present. With certain specialized exceptions,

---

[65] This discussion draws on George J. Alexander, *Honesty and Competition,* Syracuse University Press 1967.

the FTC is alone in policing the ethics of advertising.[66] It is most instructive, therefore, to examine what excites the wrath of the FTC, and what passes unnoticed, with the tacit implication of public approval; and the nature and severity of the wrath when provoked. Of equal importance is the context within which the FTC presumes to act, and within which it has been supported by the judiciary.

The FTC has generally attempted to protect competition from unfair restraint, and consumers from damaging misinformation about products. Since advertising has been significantly involved in the view of the FTC only in the latter case, it is consumer protection that is of interest here.[67]

In truth-in-advertising cases, the FTC asserts one kind of power. (It has limited power to seek preliminary injunctions in food, drug and cosmetic cases, but Alexander found this to be so rarely used as to be a virtual dead letter.) As a matter of general practice, the FTC identifies violations and issues cease and desist orders. Violations of these orders are called to the attention of the Justice Department. Any advertiser who complies with the Commission's order to cease the interdicted activity is not subject to further penalty. This lack of penalty means that the only risks an advertiser runs in engaging in falsehood are 1) the possibility of losing his investment in the development of a campaign which may have to be terminated while still potent; and 2) the possibility of purchasers using the FTC finding of falsity in pressing product liability litigation.

The first is probably the more effective constraint, since the latter rarely arises for most products, and can be handled

---

[66] The Food and Drug Administration acts in certain circumstances regarding labeling; State governments are relied on to police the wording of insurance policies; but the FTC is so overwhelmingly dominant that its actions are assumed to set the tone and to represent the whole of public monitoring of private advertising.

[67] The FTC has never seen fit to assess the effect of advertising on competition, in the sense that advertising questions might be at issue in merger cases. This may be a serious bit of myopia.

relatively easily with an in-house legal staff and liability insurance. There is little danger of damage to the corporate image since the public expects falsity of one kind or another in any case. Nor is the public likely to hear about it, since it would have to be reported in the media that accepted the falsity in the first place.

The pressure inherent in this regulation is so minimal that it would be unsafe to assign any cause-effect relationship to FTC-advertiser policies. However, if viewed as a symbiosis, advertiser-FTC relations are admirably suited to protect the freedom of action of the corporate managers of the tools. Major national advertising effort simply fits into those areas into which the FTC has never ventured. The Commission has attempted to protect the consumer from *misinformation*, and has never taken under advisement cases on the grounds of being merely *non-informative*. If an advertiser gives *some* information, his ad may be subject to interdiction for deceptive incompleteness or inaccuracy. If he attempts no information at all, he is beyond FTC scrutiny. Ads that contain information about some topic other than the product are equally immune, since the FTC would not see this as pertinent to the question of deceiving the purchaser about the product.

Of most crucial importance is a side effect of the policy. FTC scrutiny of advertising only for misinformation about the product, and refusal to become concerned with broad societal questions relating to advertising practices, constitutes a grant of almost total freedom of action in the area of political propagandizing under the guise of product promotion. FTC abandonment of this tunnel vision would pollute the corporate environment beyond endurance. Consider the prolonged television silences that would result from interdiction of the irrelevant. Or consider the difficulties of admen trying to write something actually informative about two brands of detergent or the relative services of two banks.

This constraint upon constraint is progressively tightened as corporate function becomes more highly technical and dependent upon sophisticated specialized knowledge. The tendency is, then, for government regulatory personnel to be drawn from areas closely allied to the regulated function. The greater the specialization needed, the thinner the ranks of personnel to which regulation can be safely trusted, and the less likelihood of divergence of view between regulated and regulator. Hence the commonplace observation that regulatory agencies tend to become advocates of the industries placed under their scrutiny. Whether the need for technical expertise for regulation in the public interest is real or specious is of little moment if the public can be made to believe it is real.

It is this aspect of corporate yearnings for freedom that lies closest to the center of the whole system of corporate ordering of technology. The gross national product, viewed together with the present state of tooling, illustrates the point. The GNP consists of the sum of all transactions of record. Another way of saying it is that it is the sum of all marketed busyness. To engage in and market the busyness of which their tools make them presently capable; and of which anticipated elaboration into different tools will make them capable; this is the essential freedom. The GNP faithfully reflects this. It takes no notice of whether the busyness is constructive or destructive, enriching or impoverishing, from the view of any human value. It only reflects its magnitude. And to use their tooling capability to increase that magnitude is the one indispensable corporate freedom. To this end markets are administered; failure is hedged against; size is sought, not only for efficiency but to cushion shock; and above all, the myth is fostered as truth that wisdom is technological.

Freedom to set wages may be sacrificed if rising wages do not affect the earnings necessary further to elaborate tools. Indeed a tax, which may appear to be a constraint, may provide an occasion for growth in the presence of administered prices. The laying of a tax in such circumstances simply amounts to the government contracting out its tax collecting functions as it might any other task. Whole batteries of computer technicians and tax lawyers may join the enterprise, financed by the consumer and adding materially to the total stock of present and potential tools and busyness.

I asserted earlier that corporate power in the modern day flows in large part from the fluidity of identity afforded by legal but not biological parenthood. In order for this assertion to stand, it must be shown that laws conscientiously designed to constrain do not constrain; and/or that corporations are able to prevent design of laws that would constrain in such fashion as to impair critical areas of freedom. It should also be kept in mind that "law" in this context refers to ordinary, visible, statutory and regulatory law of government. (It will subsequently be argued that corporations do indeed submit to the constraints of law, but law generated within their own peculiar legal system that makes use of, but is not subject to, the statutory laws of nations.)

Formal legal hazards to corporate freedom of action take, of course, many forms: taxation; regulation by government agencies under statutory guidelines; enforcement of general criminal statutes; the use of the governmental judiciary by the public to collect damages, as in the field of product liability, in addition to all the diffuse liability hazards of any citizen; and enforcement of those criminal statutes such as the Clayton and Sherman Acts which are specifically designed to restrain corporate behavior.

In addition to these formal restraints is the hazard of public disfavor, which might lead to the enactment of additional and unpredictable restraints. The magnitude of lobbying efforts at all legislative centers, plus public relations efforts directed toward the general public, would seem to indicate a high degree of concern for this aspect of the corporate future.

What kinds of sanction can, as a practical matter, be brought to bear on a corporation? These are severely limited. Fines or damages may be collected. Injunctions may issue, proscribing certain behavior. But what else? Imprisonment? Hardly. Imprison a document? As in the last General Electric conspiracy case (there have been more than 12), officers may be imprisoned. But that hardly touches the corporation. A host of hot-eyed young executives stand behind such persons, ready to fill their shoes. Focusing public disfavor, inducing public non-patronage, might be a real danger. A judicial order to do something distasteful such as divest the corporate structure of a subsidiary or group of subsidiaries, is possible. The possible sanctions can, then, be summed up for all practical purposes as these: the corporation may be ordered to do something (such as divestiture in a successful anti-trust prosecution); or to stop doing something (as in false advertising interdicted by the Federal Trade Commission); or to pay money (as in case of a fine or successful damage suit).

Such sanctions as these may fall with great severity upon small firms. Fines and damages tend in most cases to be pre-established in relation to an offense, not to the size of the transgressor.[68] Thus it may be confiscatory to a small firm not administering its market. The same fine would be not only insignificant to a large firm, but in an administered market could be passed along to consumers. The same holds true with damage claims. But these are not the key questions.

Earl Latham points out that regulation is always a compromise with some aggrieved public not represented within the corporate power system. Such compromises are rarely more than minimally and sporadically effective as restraints on what is to all practical purposes absolute power to legislate within the sphere of corporate interest.[69] This was

---

[68] Earl Latham, "The Body Politic of the Corporation." Mason, *op. cit.* p. 236
[69] *Ibid.*, p. 228.

included in a general description of the corporation as an analogue of the state, viewing it as a private government. This view of the corporation is shared by a number of its students, and will be examined specifically below. I cite it here in support of the contention that the mature corporation is, in areas critical to its own interests, free of effective constraints by the organs of conventional civil governments.

The phrase "in areas critical to its own interests" is so central to the question of corporate autonomy that it needs endless repetition. Most students looking for governmental power in the corporation are concerned with its relations to some public or other. I am more concerned, as I believe the corporation to be, with its relations to the tools. If Fuller, Galbraith and Ellul are correct in their belief in the autonomy of expertise or technique, then freedom of action in relation to orderly elaboration and amortization of tool systems is the indispensable freedom which may not be surrendered. All other freedoms must be viewed as derivative of this. Any which does not impinge on this may be traded off for a freedom that does so impinge.

Corporations were never planned to become sovereign. Whatever of sovereignty they possess has come to them through evolutionary accretion. Originally private institutions created by the state to conduct specific elements of public business, they had, by the end of the 19th century, become private agencies utilizing the power of the state for the conduct of private business. This is a profound change.

As an arm of the state, a firm could lose its grant of power. If the corporation becomes sovereign shortly after it comes into existence, it is an institution that defies conventional description. The exact metamorphosis that takes place following the issuance of a charter today badly needs definition. Unfortunately, precision of analysis is made difficult by the negative nature of much of the evidence.

Firms conventionally identified as American are preeminent in size in all fields: manufacturing, utilities and finance. In terms of revenues or assets, one must go down the list a goodly distance before encountering a "foreign" corporation. (In manufacturing, for instance, Royal Dutch Shell, number one foreign, is the approximate equivalent of Ford Motor Company, number eight among Americans).[70] If one is dissatisfied with explanations of this phenomenon which turn on some American genetic genius, or corporate pituitary failure, other choices are limited. Perhaps "free enterprise" in the United States has been uniquely free in those areas that permit most rapid and uninhibited elaboration of tools. It may very well be, in fact, that James Madison, out of fear of unbridled corporate power, combined with failure to anticipate the career of Chief Justice John Marshall, laid the first solid foundation for the future gigantism of American firms.

Although accounts of the debates in the Constitutional Convention omit any reference to the power to issue corporate charters,[71] Madison is known to have felt that no government should have this power.[72] Since nowhere in the Constitution is it mentioned, this became a part of the undefined residual powers devolving on the States. Though Marshall upheld the power of Congress to create corporations (*McCulloch vs. Maryland*, 1819), Congress was

---

[70] "The Fortune Directory of the 200 largest Industrials Outside the U.S.," *Fortune,* Sept . 15, 1968, p. 130 ff.   "The Fortune Directory of the 500 Largest U.S. Industrial Corporations," *Fortune*, June 15, 1968, p. 186 ff.

[71] Carl Van Doren, *The Great Rehearsal*, The Viking Press, New York, 1948.

[72] Gaillard Hunt (ed.), *Writings of James Madison*, G.P. Putnam's Sons, 1910, Vol. IX. P.281.

In a letter to J. K. Paulding Madison wrote in1827:

"Incorporated Companies with proper limitations and guards, may in particular cases, be useful; but they are at least a necessary evil. Monopolies and perpetuities are objects of just abhorrence. The former are unjust to the existing, the latter usurpations on the rights of future generations."

Particularly interesting is Madison's concern with "perpetuities," the immortality inherent in legal persons.

hedged about as in the applications of other implied powers. Close relationship had to be shown between the proposed corporation and a legitimate public function of Congress. No such constraints fell upon the States, however. The power was theirs, and by 1890 the limitations on size and nature of business which might apply for a charter had fallen away. The charter soon became, in many States, available to a petitioner as a matter of right.

The emergence of a subordinate government body, which conducts, for example, no foreign relations, as the source of corporate existence, has important ramifications. Earliest to appear was the possibility of small States using the charter power as a bid for wealth. Since corporate law could take as many forms as there were States, and since issuing a charter would almost guarantee a home office located in the State, small States were in possession of a weapon that could help them overcome the advantage of populous States with large urban centers. New Jersey, Connecticut, Nevada, West Virginia and others have pursued this course. Nevada used the power of the State over divorce to seek unique advantage. New Jersey did the same in the issuance of holding company charters. The principle is the same.

Another ramification, and one of more fundamental import, is the fact that the national government is excluded from exercising jurisdiction at the wellspring of the corporate existence. If the charter is issued by the State of Delaware, Congress cannot terminate it. While Congress may regulate interstate activities, it cannot reach the source. Thus only the State which issues the grant of its power can in theory take it back. But the power, once granted, has all the qualities that it might have if granted by Congress. The constitutional principles of "full faith and credit" and equal protection of the laws turn what is *de jure* a grant of State power into a *de facto* grant of national power. The corporate grant is made relatively secure by the fact that no agency can reclaim it except the one that granted it in the first place. Corporate growth and elaboration then make it virtually impregnable. The great firms become many times as wealthy as the States from which their charters were drawn. In addition, they have numerous other charters, drawing power not only from other States of the Union, but from other nations. How then can their sovereignty be touched? Hence the negative evidence. None of the great American firms has been threatened with withdrawal of its grant of State power. It is difficult to conceive of a circumstance in which this might be undertaken, and even more difficult to conceive of one under which it might succeed.

This peculiar legacy of classical federal theory in the United States, it should be noted, is available to foreign-based firms also. They may partake of its blessing by acquiring or forming an American-based subsidiary, with a charter from one of the several States.

One of the more quaint and charming illustrations of the options open to corporations by U.S. law involves COMSAT. The Communications Satellite Corporation, created as a private corporation by Congressional statute, promptly created three wholly-owned subsidiaries of the same name with charters drawn from Maine, Delaware and Alabama. The parent company is required by its charter to include in its board of directors three (of fifteen) members appointed by the President of the United States.[73] It is intriguing to speculate on the consequences of a Congressional decision to withdraw its charter. Would COMSAT continue to exist in the subsidiaries' identities? Might the net result be simply to remove the necessity for representatives of the public on the boards of directors of the remaining entities?

Or, in another format, consider New Jersey Standard, a worldwide corporation, with subsidiary units chartered by numerous countries. Of what significance to such a corporation would be cancellation of its parent company charter by the State of New Jersey? And what sort of flight

---

[73] *Moody's Utilities*. See sections on COMSAT and AT&T.

of quixotic derring-do would be required of the New Jersey legislature to contemplate such a move?

Each of the above cases seems to indicate a strong likelihood that corporations, once they have distributed their identities through a variety of subsidiaries, enjoying grants of power through charters from several different political bodies, are immune from destruction. Even nationalization of assets, in the case of a multi-national firm, commonly amounts to cutting an arm from a squid which is quite capable of growing it back. Nationalization would fall only on the chartered segment accessible to the government involved. The absence of actual cases of dissolution[74] of great modern corporations tends to support the notion of immunity.

Corporations are truly remarkable for their recidivist lawlessness. To find a rate of criminality among natural persons commensurate with that among the largest fifty U.S. corporations, one would have to look inside a prison. Corwin Edwards found in 1956 a close relationship between size and number of criminal convictions. General Electric had by then been convicted, mostly of criminal conspiracy, 12 times. Of the largest 50 firms, only five had never been convicted of criminal offense, and they had registered 102 convictions among them. Of the largest, Edwards observed that abuse of criminal statutes was part of routine operations.[75] And most of these firms remain in the inner councils of military technology, trusted stewards of secrets not allowed exposure even to members of Congress.

If the great corporations are possessed of sovereignty; and if they can at least determine the conditions under which they are vulnerable to statutory law; then their position with respect to national boundaries becomes an open question. The notion of "citizen" and its antithesis, "alien," are specific

---

[74] In contrast to divestiture, which does not touch the charter.

[75] Corwin D. Edwards, *Big Business and the Policy of Competition.* Press of Western Reserve University, Cleveland, 1956, and appendix.

in their assumptions of the possession of bodies and nervous systems. Privileges turn upon these categories of status, but so do responsibilities, responsibilities that are only credible in relation to natural persons. Fealty to a single political regime; availability for military service as a matter of duty and for subsistence pay; adherence to the laws of government on pain of punishment which may be permanent or lethal; in short, a citizen is expected to live by an institutionalized ethnocentricity. Corporations must exhibit fealty to every political regime that has issued a charter, even if two are at war. Corporations engage in military affairs on their own terms and in their own good time, commonly for extraordinary profit and advantage. Even more telling, they may serve many military masters simultaneously, even hostile ones.

The inescapable conclusion seems to be that corporate citizenship is a state of convenience, which can claim perquisites of natural persons but escape the responsibilities. As a Frenchman it would take me five years to become an American citizen. As a French-based corporation, I could do it by purchasing a controlling interest in an American-based firm. American citizen managers would give credibility to its American-ness, but it would now be expressing the judgments and interests of an agency out of the jurisdiction of American law.

If these were the only limitations on the fit of corporations to the system of national law, they might be considered simply mercenary citizens, to be used by any society to further its ends. But there are indications that the interests of corporations may run quite counter to those of a national society and be enforced against it, whether from a domestic or foreign base. The corporate device for doing this across national boundaries is international public relations and is available to any agency that will pay. Perhaps the least publicized of all major corporate undertakings, this permeates the public information streams of the contemporary world. I have seen it explicitly described in only two places.[76] Briefly it works like this: one member of a public relations, advertising or legal firm registers with the State Department as an agent of a foreign power. That firm then may accept retainers from foreign governments or other agencies to propagandize in their interests. Such propaganda activities derive their value from the obscurity of their source, hence the extraordinarily low visibility of the whole system. Elaborations are, of course, legion. A candidate for public office, no matter how obscure, has no way of knowing whether his opposition is being financed by such a means, with the impetus coming from a foreign government or other extra-national agency. Lobbyists in Congress and in State legislatures may be so employed. What additional freedoms are available through the use of subsidiaries is not clear. Several large agencies maintain subsidiaries based in foreign countries, and grouped for management purposes into the typical holding company. Since few such firms are publicly held, reports of billings are neither complete nor readily accessible. There is nothing detectable in the system to prevent truly hair-raising combinations of tasks. A given firm might, for example, conduct simultaneous propaganda campaigns for the Air Force (at the taxpayers' expense); for a petroleum firm in its negotiations with a foreign supplier of crude; for that foreign state, in pursuit of its foreign policy interests in the United States; for a presidential candidate in his bid for election (in any number of states); and for any number of private agencies, domestic and foreign.

Viewed in conventional light, the conflicts of interest inherent in such maneuvers are simply staggering. The conventional light probably illumines very little of this, however. This aspect of corporate activity, like all the rest, is cited not to demonstrate evil, but to indicate corporate

---

[76] Drew Pearson, *The Case Against Congress,* Simon and Schuster, New York, 1968, Chapter 13. And Douglass Cater and Walter Pincus, "The Foreign Legion of American Public Relations," *The Reporter*, Dec. 22, 1960., pp. 15-22.

transcendence of conventional social institutions and restraints. The legitimate acceptance of money for the performance of duties clearly seditious if carried out by a natural person, most pointedly illustrates the impotence of national boundaries in the corporate world. It also indicates rather dramatically that no nation's domestic politics is necessarily what the various publics think it is. International rules against one nation or person tampering with the internal political processes of another nation are readily and invisibly bypassed.

If the corporate world is one in which one agency, with the legal status of personhood, may simultaneously work to arm two states, even if they are hostile; may promote for pay in one state the propaganda interests of another; may work for the election or defeat of political candidates at the behest of foreign governments or private interests; may escape the constraints of national law, and remain unharmed by any number of criminal convictions; may capture and staff the regulatory agencies created to protect the public interests, thereby making the public interest over into its own; and at the same time increase its already overwhelming control over the world's work capability, including that of informing the world about its own activities; then the corporate world is not one that is comprehended by the natural persons who are its beneficiaries and victims. Neither the language of politics nor of economics explains this world. Indeed the fact that they are separate languages helps conceal it. The only conflicts of interest that are relevant are conflicts with the imperatives of tools.

But we are natural men, living in a natural world of oxygen and photosynthesis, microbes and fishes. If we do not come soon to comprehend the mechanics of our energy-controlling social system, we will only experience our fate, not understand or control it.

Chapter VII

TRANSCENDENT ORDER

Hail and beware the dead
who will talk life
until you are blue
in the face.
And you will not understand
what is wrong...
Charles Olson

A host of individual sovereign units, amorphous in form, uncontrolled by conventional politics, and impelled in their careers by the inexorable necessity to amortize and elaborate their tool systems, seems at best an incomplete picture. How are the great corporations related to each other? By the market? Surely one need not inter Adam Smith and the sovereign consumer again. But to deny that corporations are functionally related is to deny the world around us. The copper from the giant mining concern finds its way with apparent smooth facility into the transmission lines of the great utility companies. The computers of IBM and the copy machines of Xerox find their ways into their relevant slots in thousands of different operations with a minimum of controversy. The graduates of the great (and small) universities seem to find ready employment on the technician-hungry corporate payrolls. Freeways unroll in front of multiplying swarms of automobiles. Thousands of subcontracting corporations pool their talents with the prime contractor in the development of a single weapon system. Rand Corporation theory games generate strategic notions which, fed through government, surface as nuclear deterrent and return to corporations infinite possibilities of foundering the continent under the weight of missiles. Manufacturers, sellers, advertisers, media, managers of money, dovetail their worlds in ways that undeniable work. If the corporate units are not what convention would have them; and if they are related one to another in such obviously successful fashion; then perhaps the framework of their togetherness is outside of convention also. Let us return to the tools for clues.

If Fuller's conception of the tools of the industrial phenomenon has validity (a network of tools, growing on its edges out of the current generation of tools), then the institutional control over that network must respond to its imperatives and ideally to no others. There are certain properties that are denied to controlling institutions by the requirements of the tools. Perhaps the search for what *is* can be narrowed by the exclusion of what *cannot be*, and still allow the mechanical heart to function.

The system of control cannot be anarchic, and it cannot be sharply subdivided. Mechanically, no segment can stand by itself or come into being in isolation from others. Start anywhere in the system and trace its ramifications. The functional unity is inescapable. Not only must General Motors have access to raw materials and transport (for which it must depend on other units of the system), but such access must be orderly and predictable on a long-term basis. Long lead times and heavy advance capital commitments lend increasing urgency to these requirements. Access to media for dissemination of vital propaganda, both product-oriented and political, and transport for distribution of product are equally imperative. None of the myriad intertwined streams may be allowed to fail. Credit, insurance, banking, mining, research, all must function smoothly. Each individual segment is dependent on the whole, and each is subject to the imperatives of specialized tooling. Industrial anarchy is inconceivable. The existence of each and every industrial good gives eloquent witness to order. Knowledge must be moved and withheld on schedule, as must goods. New tools must be incorporated into the art, but neither too soon nor too late.[77] The teaming of presumed competitors in confronting challenges (autos before Ralph Nader, or chemicals before Rachel Carson) is a highly visible example of the presence of order, the absence of anarchy. Advertising campaigns mounted by industry-wide associations testify to the presence of common interests, but more cogently to the institutional mechanism for their orderly pursuit.

The asserted impossibility of sharp subdivision is more difficult to demonstrate than the impossibility of anarchy, though in a sense it is the same problem in a larger frame of reference. One would expect sharp division between railroads and autos, for example. But at any given time the milieu of each includes the other. It is to everyone's interest to dwell symbiotically rather than at war. So the auto industry fights for a growing share of passenger movement; and the railroads fight to free themselves of passengers, limiting their national advertising to their movement of freight. The success of each in bringing order to its own world is important to the other. One brings raw materials to the other, while enjoying a market for its capabilities in the busyness of the other.

Similarly, subdivision on a greater scale is hardly possible. The world's universities are full of foreign students and scholars. Viet Cong bullets have similar muzzle velocities to American bullets, hardly a phenomenon of

---

[77] For the story of the letter patent, by which the rate of change of industrial tooling is controlled, see Mason, *op. cit.,* pp. 154-164; H.L.Nieburg, *op. cit.,* Chapter XV, "Patents and Power"; and Walton Hamilton *op. cit.,* "The patent system in action," pp. 78-92.

These authors make it imminently clear that, as in other aspects of industrial society, the institution of the patent is not as it seems, and not as it is commonly said to be. Designed to protect the individual inventor, it has become a major tool of corporate power in defense of the essential freedom of action, and in promoting corporate gigantism. Firms stockpile patents to maintain single-source status in military procurement. They use their laboratories to "invent around" patents held by individuals, etc.

independent invention. The Russian and Chinese intercontinental ballistic missiles can be alternatively invoked as justifications for the same ABM installations. Their technical similarities make a *prima facie* case for the freedom of technological movement across the most divisive political boundaries. Industrial tooling, wherever it occurs, clearly grows out of existing tools, conventionally conceived barriers notwithstanding.

Let us hypothesize a society of legal persons, evolved along the course that has brought us from Gutenburg and sailing ships to TV dinners and the moon. Able to escape or avoid at will the constraints of national law and international boundaries, it is governed by its own internal legal system of treaty and contract. Central to its purpose is the implementation of its own ethical system and avoidance of the ethics of natural persons.

Hence it exists largely in secret, revealing to the natural public only those aspects of itself for which general acceptance is anticipated, or that are necessary for the manipulation of the natural populace to its ends. To the secrecy available to the natural person through private agreement and contract, the corporate society adds the privacy afforded by the preeminence of these modes of conduct. While the natural person's privacy is always hostage to his compliance with a public statutory law, the privacy of the corporation is relatively inviolable. Statutory laws can command small attention from the population of a society of corporations.

A hypothetical society of corporations is difficult to talk about, because the rhetoric of the nation state is all we have. National comparisons are made of steel production, income, energy consumption, gross national product, or any of a host of other statistical values. These are taken completely at face value. But it may very well be that the national societies of natural persons are collecting the data but not doing the work. It may be that a single Fullerian worldwide network of energy flow, generated by and channeled through a single worldwide network of tools, is managed by a single worldwide system of controls quite separate from the societies of natural persons. If this is the case, the comparisons of national values relate to work done *within the national boundaries*, but not *by the national society*. The function of national divisions, then, becomes simply a reflection of the agencies collecting and reporting data, not the agencies generating the phenomena measured. The work is being done by some agency or agencies whose extent and authority are consonant with the task of managing a worldwide network of tools. No national agency meets this requirement. Indeed no agency of natural persons meets this requirement, since all are subject to control of their respective political states, none of whose boundaries are coterminous with the industrial phenomenon.

A non-national corporate society, if it has substance, is well concealed by this adherence to the national boundary system. To the extent that it is used in describing the world, it denies without refuting the existence of another kind of world order. Corporations have populations that relate to them, but they have no boundaries that can be geographically defined from public information. It may be most useful to assume that they have no boundaries at all. Their worlds at any given time include not only all those people and firms with which they have transactions, but all those with whom they aspire to establish transactions. At any given moment, a corporate message is reaching a particular population through the media. At another moment, the population is quite different. *All* such populations are part of the corporate world. To induce patronage, to create a favorable public image, to abort dangerous political developments, the corporations reach out to anyone reachable at the moment. This defies conventional definition. Political boundaries, or any geographically defined boundaries, have no more relevance in this milieu than they have to radio waves. That the tools work seems the best evidence that a system of organization is emerging that is derivative of their properties.

Corporate capacity to transcend national boundaries in organization is matched in operation. A corporation whose parent charter is drawn from state A can operate in state B free of statutory constraints of either. Consider the case of the government of state A (as in the case of the United States in 1968), out of concern for its balance of payments, placing restrictions on the movement of capital to state B and requiring the repatriation of profits from subsidiaries located in state B. J.J. Servan-Schreiber[78] testifies to the futility of such legislation, as it might be undertaken by either state. He points out what happens when the subsidiary's host country (state B) attempts statutory restrictions.

> When an American corporation decides to cross the Atlantic and set up shop in Europe, it doesn't much care where it locates its plant. It can, so to speak, put itself up for auction among the competing European governments to make the best deal. And it gets what it wants.

Not only does this indicate a primacy of corporate freedom over national freedom; but suggests that it is now the public political units which are in competition, rather than the private industrial ones. Adam Smith would choke.
Servan-Schreiber again:

> A Common Market country that takes a more restrictive attitude than its partners toward American investment only helps its competitors at its own expense.[79]

Thus the state becomes a tool of the corporation, something to play off one against another to increase corporate freedom and promote corporate growth. There is

---

[78] *The American Challenge,* Atheneum, New York, 1968 (first printed as Le Défi Améeicain. Editions Denoel, 1967), p. 16.
[79] *Ibid.,* p. 19.

another and more subtle aspect to the phenomenon alarming Servan-Schreiber, illustrated by the following passage. After describing gross percentages of American control of French industry, he says:

> The most important sector of the economy, however, and the one most crucial for the future, is electronics. Here it is easy to see the direct link between the role of American firms and a high degree of technology involved in production. American corporations in Europe control:
> 15% of the production of consumer goods—radio, TV, recording devices, etc.;
> 50% of semi-conductors—which now replace electronic tubes;
> 80% of computers—high-speed electronic calculators that, among other things, now even transform the management of corporations.
> 95% of the new market for integrated circuits—miniature units crucial to guided missiles and the new generation of computers.[80]

He is exactly correct in seeing the gradation in these figures as crucial. The progression is toward the evolving edge of the network of tools. But suppose one remembers that the industrial giants of the corporate world are predominantly thought of conventionally as American, and then abandon the notion of nationality. Then what Servan-Schreiber is seeing is the capture by the corporate giants in the worldwide corporate society of the crucial aspects of the tooling, wherever they exist or can be developed. This fits impeccably with the notion of the corporate society. Servan-Schreiber, like the rest of us, is viewing the world from the preconceptions of the nation state. He is worried about the preeminence of one state over another. He is a Frenchman

---

[80] *Ibid.* pp. 12-13.

and a European, and is looking for ways by which Europe can compete with the United States. This is a pure nationalistic view, though convention would call Servan-Schreiber an internationalist. The nation to which he is committed is admittedly Europe, instead of any existing state, but only the boundaries have changed. The view is nationalistic.

If members of the hypothetical society of corporations transcend and use nations, one must abandon national identification of firms. IBM, the computer giant of the world, is expanding its capabilities by obtaining control of and transforming existing firms across the Atlantic. This is happening. Bringing national names into the discussion may obscure rather than reveal what is going on.

Since it can escape the full force of statutory law of nations, this society is not defined by national boundaries except in special narrow circumstances. Technological and institutional operations within the purview of this non-national society of corporations being constantly and profoundly inimical in certain respects to the interests and value systems of natural persons, a constant screen of secrecy must be maintained. This screen has evolved quite naturally, as more and more of the processes of generating and disseminating information have come to depend on industrial tools, and hence have moved within the scope of control of mature corporations. One of the most fascinating aspects of this screen of secrecy is that it must conceal, among other things, its own existence. At all points it must stand between what is going on, and what the natural publics think is going on. It must constantly present information in such a fashion that the natural publics will not be led to try to interfere in any disruptive way with corporate freedom of action.

This screen of secrecy has two major components. Generally it might be divided into what is clearly secret, deriving from the private nature of the corporation and its operation through treaty and contract; and what might better be called illusion. The latter is achieved through corporate control of the bulk of the total flow of information to the natural publics. This should not be thought of as censorship, but as the consequences of routine judgment in the design of the contents of the mass media. The media, of course, exist as systems of industrial tools, the control and ownership of which is lodged in the society of corporations.

The secrecy generally breaks down when the contract system fails, and brings into corporate affairs the investigative and judicial organs of government, armed with subpoenas. Illusion can be pierced by careful work, and often is by scholars or journalists or other inquisitive types. This rarely matters, however. Even if such works are widely read, which is rare, the weight of the endless stream of counter-information pouring from the schools and other mass media is sufficiently overwhelming to make them impotent. The willingness of publishers whose houses are part of corporate conglomerates to publish books such as those reviewed at the beginning of this one, bears witness to the mildness of their threat to the screen. Those who read and are deeply affected by such books are apparently so few that they form no effective public opinion, only a modestly attractive market.

The television rating systems are well known, and their shortcomings hardly need additional review. They do, however, present a highly visible model of a ubiquitous technique in maintaining the screen of illusion. The technique is to declare the intent to bring the public what it wants; present the public with alternatives all of which are acceptable to the wisdom of the tools; measure public reaction to the alternatives presented; and declare that the one which sells best is in fact what the public wants. Though it is less visible in the field of auto manufacture, textbook publishing, detergents and light bulbs than in television programming, the system is no less present. Inherent in each situation is artificial product differentiation through advertising or other forms of information dispensation, to create the illusions of a broader range of alternatives than actually exists, permitting more exact adherence to the wisdom of the tools.

The hypothetical non-national corporate society is not something that can be clearly shown to exist. It is a way of looking at a phenomenon that has come into being gradually, much of its evolution having been concealed by increasingly inapplicable language of social description, such as the formal premises of politics and economics. The population of great corporations today is the least understood and least studied aspect of social organization, while holding a position of absolutely unprecedented power over the human scene. The concept of a corporate society whose population is made up of mature corporations provides a frame of reference that appears capable of subsuming those things that empirically can be observed as happening. The implementation of an ethic derived from the imperatives of tools, no matter how this might conflict or accord with ordinary human values, and the shaping of the language to project the necessary rationalization; the conduct of most work in secret from the public it is presumed to serve; the freedom of the agents in possession of the secrets and the tools to transcend the constraints of statutory law, which often embodies ordinary human, rather than technological, concerns; the ability of these agents to move at will across national boundaries, even to the extent of promoting simultaneously the interests of hostile nations; their ability to define all problems requiring solutions as technical problems, subject to technical expertise, no matter how subjective and value-loaded the problems might appear if viewed from a non-technological perspective; their ability either to immobilize or pull into their own orbit institutions whose historical functions have been antithetical to the wisdom of the tools, institutions such as universities, churches, and government; the ability to use any technological threat to the health and life of natural humans as an excuse further to escalate the application of energy which creates the threat; the conventional notions of political and economic institutions hardly identify, much less explain, these characteristics of our contemporary existence. As I am writing this, NASA's astronauts have just landed on the moon. The television networks are trying to produce a thirty-hour spectacular, and are at least providing an enormous audience for their commercial sponsors. I am experiencing a disturbing feeling of simulation about the whole thing. This, after all, is the first of man's great feats of exploration in which the presence of real humans is simply an additional technical problem to be overcome. According to newsmen's interpretation of what was going on, special equipment had been installed to allow the crew to interfere, if they desired, with the fully computer-controlled landing. One is tempted to the suspicion that the heroic presence of Armstrong and Aldrin has value principally in strengthening the screen of illusion that rationalized the undertaking. Something like $25 billion is reputedly the government tax cost of getting to the moon. Equally illustrative of the whole phenomenon, however, is that the worldwide television transmission facilities, reaching hundreds of millions of people, was used to transmit the judgment of an unidentified young man in Manhattan that "now we can solve the population problem." The cost of this transmission will show up in the retail prices of the sponsors' products, book-kept at something like $100,000 per minute. The task media technicians have apparently set for themselves is to produce a continuous hymn of excitement and awe, extracted from the technically uninformed or from those whose careers are an integral part of the technical effort. The result is an extravaganza in which the role of natural humans is hardly more than simulated. The event is endlessly touted as mankind's greatest feat, but reasons for this judgment remain obscure. They must remain obscure because they have meaning primarily in the framework of tool elaboration. Apollo XI is certainly man's most stunning example of technical sophistication. Judged in terms of the interests of natural humans instead of legal humans, it is hard to see it as other than a pseudo-event.

Looking to the future, a NASA official remarked in an interview just after the landing that two major questions remain to be answered: Can a degree of cost attractiveness and reliability be developed in space travel commensurate with that now enjoyed in air travel; and second, once that is accomplished, will people want to travel in space? Produce it for sale, and see if anyone can be convinced to buy. Shades of the Nielson ratings. The advertising campaign necessary to pay for amortization of the generations of tools following those of Apollo XI boggles the mind.

Rejection of the notion of conspiracy is absolutely central to the entire argument. In the face of a divided culture, with technology changing more rapidly than the capacity of value systems to judge and legitimatize it, people have tried to find a course both operational in relation to the new tools, and reasonably comfortable to mind and spirit. Separation of function—between economics and politics—came first. When this ceased to be adequate, separation of person followed, with the elaboration of the idea of the corporation. Thus human characteristics can be retained in natural persons, and the characteristics required by the technology and its theory are assigned to legal persons. This has not been accomplished consciously, or it would have been self-defeating. Rather it has been accrued through small steps, taken each in response to the situation of the moment and the postulates in use at the time to rationalize it.

Chapter VIII

CLOSURE

and the slithy toves
Did gyre and gimble in the wabe…
Lewis Carroll

The nautical notion of the "steady bearing' is really a delightful bit of Euclidean simplicity. Let us say that vessel A is traveling a compass course on the open sea. Vessel B appears on the horizon due east of vessel A. The exact direction is determined by taking a bearing. Fifteen minutes later another bearing is taken and the direction has remained unchanged. Vessel B is still due east. The situation is now clear: vessel B is getting closer (if it were not, it would not have appeared over the horizon at all); and somewhere ahead there is a point at which the vessels will collide. The data of the bearing are not sufficient to determine how far ahead that point is, but make its presence at some distance a certainty.

Let us assume that the course of vessel B cannot be altered, nor can its speed. The avoidance of collision, then, is entirely the responsibility of vessel A. Vessel A has several choices: if it increases its speed it will pass ahead of vessel B; if it slows down, it will pass behind; if it turns toward, it will pass behind; if it turns away, it will either pass in front or not pass at all. *Doing nothing guarantees eventual collision.*

Vessel A is taken to represent the exponentially increasing work capability of industrial tools, together with their managing institutions and supporting mythology; vessel B is taken to represent the tenuous organic linkages that make life, including human life, possible on earth; and the collision represents some set of circumstances, unpredictable in time but more and more certain in the foreseeable future, which will terminate the human species. Whether the collision is likely to take the form of warfare; environmental poisoning; interruption of the oxygen, phosphate or nitrogen cycles; famine; or some other creeping cataclysm; is a cosmic quibble. In the larger frame, it is all of a piece: the end product of the industrial system; the asymptote of the exponential curve of rising industrial capability.

The purpose of the nautical model is to illustrate the dilemma in its simplest form. We are the passenger-owners of vessel A. We have hired the crew, but don't know who they are. We can see that not only is the course not being changed, the steering mechanism is being dismantled. In response, we are deciding geometry is controversial and are redecorating the ship's lounge for a party in honor of arithmetic. In the faint sounds of warnings of the continued steady bearing, we find a rhythm to which we can dance.

Let there be no mistake. The wisdom of the tools must ultimately be lethal to men. This is a certainty, in the language revered by the technologists themselves, that of binary numbers. If these seers cared to program the computer to cycle and recycle the past into the future, in the game so dear to the hearts of the simulators; and if the program were designed to reflect the exponentially increasing levels of energy moving through the insane complex of present tools, the flickering ones and zeros could arrive at only one projected world, a world in which the delicate membrane of biota could have no place. Give or take a flicker, time only is in doubt. *This* is the wisdom of the tools.

The future of men lies in a subjective, non-technological, mystical if you will, but rational, refusal to be simulated. We need to assert our simple, natural, human sanity. If it is not enough, so be it.

The deadly wisdom of the tools by which we live these days is constructed of elaborate and subtle fictions. If these fictions were in the form of falsehoods perpetrated on a hoodwinked public, they would be relatively easy to combat. They are this, but they are also much more. They seem to amount to an agreement with ourselves that we will not probe deeply the semantics of public or private discourse. The effects of this on the language and its role, on all the various avenues and mechanisms of communication, are difficult to assess and describe, in part because they infect the language in which the effects must be described.

A cultural relativist might argue that all people live by such fictions; fictions, that is, as viewed by any other people. If each people lives by its fictions, i.e., its own set of myths carried in its own language, then the notion of *fiction* is transformed, and each people lives by its own truth, conveyed in its own language. In this sense and circumstance, culture and language are not separable, and the world as reflected in each culture is its real world, no matter how it conflicts in its nature with the real world of someone else's culture. Thus in any culture, individuals can use the system of their culture to *discover the nature* of their own real world, even though in a larger sense the culture *creates* the nature of their real world by providing the systems through which it is to be perceived.

The fictions of our technological society, however, are not of this sort. They are, rather, denials of the real world as discoverable by our ordinary systems, and describable in our ordinary language. Whereas language may be a medium to relate a people to each other and to their environs, we seem to be using language to create realities that our systems, applied with integrity, would deny. The white, rich, crime-free, wholly suburban society portrayed in elementary school texts; our official descriptions of Vietnam as a non-civil war; our use of the word "quality" in describing planned obsolescence in merchandise; our pretense that *laissez faire* economic language has relevance in today's monopoly marketing, wherein demand has become a manufactured good like any other; our pretense that traditional political processes still prevail when Presidential campaigns are conducted by Madison Avenue; our proclivity to speak of an infinite proliferation of doomsday devices as additions to our security; these illustrate the spectrum of our lives lived by a language carefully disconnected from the realities created by our traditional cultural systems. Daniel Boorstein has described a particularly startling aspect of this mutilation of the language.[81] He points out that advertisers have managed to redefine "truth," not just in the public vernacular but officially and legally, to refer to any statement not demonstrably false. Sequential *non sequiturs*, statements of open-ended comparison, statements based on meaningless numbers, all are accepted as "truth." They are not even susceptible to the charge of being "deceptive," much less "false." And advertisers *are* the media. That the redefinition has in fact taken place is perfectly clear from the fact that the streets are not barricaded by a public outraged by public and private rhetorical fraud that increasingly masks the basis of policy in innumerable spheres of action and inaction.

With less detachment and more passion, Goodman makes the same point in his frightening book, *Growing Up Absurd*,[82] in which he describes the predicament confronted by our bright young people who are astute enough to realize that they will not be recognized as mature until they accept lies as truth, and do proper obeisance to the virtue of basing public action on verbal fraud. As Goodman describes them, the initiation procedures of American society make those of the Sioux and Comanches seem gentle and humane by comparison. Bodily mutilation may leave less severe scars than what amounts to a contract agreement permanently to abdicate reason.

In a miasma of environmental pollutants, the very worst is the pollution of the public language. The juxtaposed words issuing from the television set, and their accompanying pictures, constitute beyond a doubt the most powerful medium of communication and diffusion of verbal, graphic and pictorial ideas ever experienced. The climax of the

---

[81] *Image, or What happened to the American Dream,* Athenum, New York, 1962.
[82] V. Gollancz, London, 1960.

Apollo Project is reputed by the media to have simultaneously reached 500 million persons, or one in seven of mankind. In what ranks with Apollo itself as one of the handful of man's most sophisticated achievements, accomplished at the cost of untold amounts of capital and, of greater moment, an enormous number of the world's most able and highly trained personnel; the MEDIA, under the control of corporate interests and with the expertise of the advertisers, regaled this giant slice of mankind with a slick and superficial hymn to the infallibility of the engineers. The public-personality newsmen expressed awe at the incredible work (technician-time) invested in the smallest detail of equipment. No relationship was suggested with the current Senate hearings on, among other things, cost overruns on government contracts. It was as though Philco-Ford, Gulf, Bendix and a plethora of other firms, from the fullness of their hearts, created the miracle for the TV audiences. Summing the time on the three networks, the public was in effect treated to 90 continuous hours of commercial, extolling the virtue of NASA and the aerospace industry. No slightest risk was run of informed or articulate criticism creeping onto the airwaves. What an appalling waste. What a massive insult to the intelligence and curiosity of the human being. What a massive abdication of simple human concern for the minds of people.

One of the most beautiful people of my acquaintance, a Methodist minister, Rev. Henry Gernhardt, explained to me one day his sense of responsibility to his audience. He said that each week, as he began to plan his sermon, he reminded himself that in talking 30 minutes to 100 people, he was thereby responsible for 50 listening hours. He had no right, he said, to waste 50 hours of peoples' time, and had jolly well better have something to say that would be worth their time to hear. If each viewer-listener tuned in on one-sixth of the total Apollo broadcast, the three major networks were responsible for something on the order of 7½ billion hours, or 856,000 years, of people's looking and listening time.

Based on the networks' estimates of their television audience, and again assuming an average viewing of one-sixth of the total, they were responsible for 325 million hours, or 37,300 years, of viewing by American people alone, a goodly congregation.

But my friend's wisdom was not that of the tools. Tools call for a different kind of approach to valuing communication. In the value system of technology, good communication is that which will elicit the desired behavior from the greatest number of the target population. "Having something worth listening to" is a nonsense notion in this context, as is the notion of informing for the sake of informing. Since the goal is to induce behavior which will support (at best) and refrain from opposing (at worst) those things the corporations are now doing (in this case enjoying immense risk-free subsidies from government in return for building and flying space vehicles); or those things the corporations will be desiring to do next (which is at present unknown); public impressions of corporate omnipotence and infallibility are the most valued consequences of media activity. Information which might serve as the basis for doubt of the value of infinitely extended and complex busyness is to be avoided at all costs. The details of the busyness, what is being eaten; how the life support systems work; what the endless acronyms stand for, etc., these are safe grist. Possible hazards can be exploited for moments of high drama, first emphasizing the possibilities of tragedy then reassuring the viewers with accounts of the incredibly elaborate steps being taken to minimize the danger. Such potential hazards during the flight of Apollo XI ranged from the possibility of engine failure in different situations, to proton bombardment from the sudden onset of solar flares. Viewers were assured that the world's radio-telescopes were keeping the sun under constant surveillance for flares, but it was left somewhat hazy just what would be done if they showed up.

One of the more intriguing aspects of the whole broadcast was the muted role of individual firms, and the blurring of just who was doing what. Individual corporate names were used from time to time, but no clear notion could be acquired from the broadcast of how many corporate participants there were; who were the prime and who were the subcontractors; what NASA's role was in relation to the private firms, whether operational or supervisory or both. Indeed, the feeling was very much that of one integrated unit, in which corporate names were largely irrelevant. With the added surrealism of television viewing, I almost had the feeling of looking at a closed community of people of a strange, homogeneous culture, who had no real reason to speak at all. Apparently blank sheets of paper endlessly shuffled across banked rows of desks; flashing lights, buttons, switches and dials seemed to take care of necessary communication. Voices seemed to serve only as reassuring noises to let those out of sight of each other know that they were standing by the correct switch or light. There was a MacLuhan-esque tribal quality, which suggested that each individual was responsible for his own portion of the ritual, the totality of which would properly propitiate the correct gods.

It may be that for thirty hours the public had the opportunity of watching in action the world's most sophisticated example of Galbraith's teams of specialist-experts whose small decisions, taken in the aggregate and passed up the chain of authority, eventually harden into the imperative of industrial processes, the wisdom of the tools, the ultimate industrial revolution.

Let me seem to digress without, I think, doing so. It is passing strange but seems to be true that when an author writes a best-selling book which in its subject matter falls within the purview of a particular academic discipline, the author loses caste in that discipline. It seems to matter little what the author has to say; the displeasure of the affected priesthood is assured by the fact of the public readership.

Why this should be I am not certain, but Galbraith, Robert Ardrey, Jomo Kenyatta and perhaps Arthur Schlesinger in recent years seem fair examples. Perhaps it is iconoclasm that offends colleagues and attracts lay readers. Perhaps there is a tendency for books to sell which expose clay feet. Or perhaps academics are well justified in discouraging those who can appeal to a lay readership, because such people may be able to persuade without a case. At any rate, it seems to be one of a number of symptoms, of a rather harsh division between the general public and the academic community (if that term can still be used, post 1968-69). This division is highly significant in the matter of communication, the public language, and the autonomy of technologic ethic and value.

The technology requires of the universities a continuous supply of specialist-experts to fill the corporate ranks. It is the constant entry of new people onto the bottom of the ladder that both creates promotions for those already employed, and brings to the cadres the latest university-designed expertise. These newcomers are the designers of the next generation of industrial tools, and the inventors and promoters of work for those tools to be put to. From the public, on the other hand, the technology requires a favorable public opinion both toward its goods and its wisdoms. For this, the tools of public relations technique have been elaborated. The task of public relations is to create and maintain favorable "images" for the agents of technology in the public eye. The concept of the "image" is to the pollution of the language as DDT is to Lake Michigan, except that its carriers, designed for public consumption and moving in interstate commerce, do not get seized by the Food and Drug Administration.

Thus the technology places very specific and compelling, but different, demands on the universities and public. University people are not numerous enough to be politically decisive in elections and in Congressional and governmental policy. They are more likely than the lay public to generate individuals who will attempt to penetrate corporate secrecy

109

and thereby threaten corporate freedom of action. The traditions clustered around the university idea of academic freedom pose a constant problem of diplomacy for the corporate ministers. Money, status, and strategic naming are the most obvious devices available to the corporations to still liberal twitching in the breasts of professors. The latter is a most versatile device. Employment becomes consultation, pay becomes the reward structure, on-the-job training leads to examination rather than to social change, since the students are there to get the expertise technologists will buy; and their writing is read by so few and excites so few of its readers that even a best-selling piece rarely causes more than a temporary ripple. In the past two decades only two books, Ralph Nader's and Rachel Carson's, have caused any publicly visible straining on the part of technologists to counter. And neither of these authors was an active academic. Nader is a lawyer, and Miss Carson was a biologist with the United States Bureau of Fish & Wildlife. The point of this brief digression is to illustrate that the corporate world has devices (tools) tailored hopefully to manage the behavior of all elements of society which in any way are, or are anticipated to become, threats to free implementation of the wisdom of the tools. Like other tools, these are always subject to elaboration, the new generation springing from the old. But all such devices are tools, techniques to be applied to elicit desired behavior or eliminate undesirable behavior on the part of the public or any of its subdivisions and adjuncts.

The language of public discourse has become, in Herbert Marcuse's words, "evocative rather than demonstrative." [83] Public language might be said to simulate society in such a fashion that society can legitimatize the simulation, that is to say, itself. When the phenomena of society become impossible to legitimatize then surrogates for the phenomena appear which can be adjudged legitimate or virtuous. It is not

---

[83] *One Dimensional Man,* Beacon Press, Boston, 1964, p. 91.

possible to legitimatize the development of virtually untreatable plague, so the surrogates of CBW, deterrent, and national security appear in its place. The substitution of "image" for description; simulation for event; name for concept; acronym for name; these are a few of the linguistic devices by which the wisdom of the tools usurps without detection the place of that wisdom which draws its meanings from the affairs of natural man.

Not all language is so infected, of course, or it would not be possible to discuss it. The portion of the language that has fallen victim is precisely the language of public discourse, the language which serves to mediate between the technology and natural men. It mediates between the closed rationality of the realm of the tools and the subjective, open rationality of politics, of valuing, of ought and ought not, of natural humanity. It is not possible to draw clear limits to the extent of the pollution, however. Since all of the population of technological societies are exposed frequently to the same messages in the language of the mass media, every person's vernacular is exposed. It is the amount of exposure to *non-technological* language that varies most. For those who read and converse widely, the pollution is dilute. For those who rely principally on the mass media, the pollution approaches saturation.

"The language" affected by this technological pollution is the-language-in-use, not the-language-as-existing. This serves further to conceal the pollution, and make it more difficult to isolate for examination. The mechanisms of pollution may be situational, as a statement that appears in a State of the Union Message, or visual, as a statement made to the accompaniment of the elaborate staging of a television commercial or news broadcast.

Since the system of tools is deemed to be the basis of survival of the society, any collision between the interests of tooling and natural humanity must ultimately be resolved in favor of the tools. The function of language is to resolve the issue in such a fashion that the interests of natural persons

111

seem to prevail, over the cold concerns of tooling, while tooling is left freedom of action; and the function of language conceals itself. Language usage designed for these purposes, then, constitutes a part of the industrial tool system, and is subject to elaboration and invention as are any other elements of the tool network.

A singularly revealing example of technological language in operation occurs in the regulatory postures of government agencies. Consider the attempts of such agencies to serve the interests of consumers by requiring specific labeling practices for food, drugs, etc. Two side effects result, which are not part of the agency plan. One is that the labeling requirement reinforces the necessity for packaging, which is a whole universe of potential busyness in itself. Lower cost bulk handling alternatives are excluded. The customer pays for the package and the label. This illustrates in another sense one of the imperatives of industrial tooling logic. It is based on the proposition that producers' responsibility is to inform the public of its hazards, not to refrain from subjecting the public to hazards. The latter would inhibit free elaboration of the tools, since certain products would be unmarketable. It has been seriously argued, and not resolved, in cases over truth in advertising, whether "stupid" consumers deserve protection from advertising fraud.[84] It is argued that, since verbal fraud is part of the game, the buyer should expect it and penetrate its workings. Thus the question is kept focused on how much fraud and of what kind, rather than on the right or wrong of engaging in any public fraud. The upshot is, of course, a gradual increase in the variety and potency of fraud. As one instance is decreed unacceptable, a dozen more are launched, some of which will survive. The "creative people" in the agencies are the specialists in this kind of tool elaboration.

The advent of television has profoundly altered the range of possibilities in communication. It has allowed

---

[84] Alexander, *op. cit.*

juxtaposition of word groups and visual images in packaged units of meaning, which serves to remove the older limitations of sense and nonsense. The importance of this cannot be overstated. It gives those who manage the content of simultaneous picture and word the power to use language through the entire spectrum, from alphabetic to ideogramic. All the communicative powers of either or both in combination may be invoked. Words may contradict images. Techniques for doing this are objects of research and development, just as in the case of hardware tools. It might be argued that printing permitted the same power, but not so. Though printing allows juxtaposition of images and words in any combination, the viewer has time to assess, with his alphabetic biases, what is going on. The fleeting images of television allow no such leisurely contemplation, and leave impressions only, repeated fleetingly but persistently.

Consider the oft-used pitch for one of the nationally advertised cigarettes. An automobile is shown caught in the press of freeway traffic. As it leaves the line on an off-ramp, we are told that this cigarette is not for everyone, just for independent people. The auto and its alluring occupants (hair halfway between executive neat and hippie free-soul), having just left the freeway, are driving in total isolation through a rural idyll.

Each constituent proposition, taken in isolation, is nonsense. But they are not taken in isolation. That leaving the freeway leads to a rural idyll; that leaving the freeway indicates an independent spirit; that a cigarette company would try to limit its market to independent spirits; each is hogwash, of course. But the whole, unanalyzed, is an ego-prop for those who smoke those cigarettes. Or the company willingness to spend tens of thousands of dollars per minute in broadcast time indicates a rather strong belief in this, at least.

Part of the mythology of the tools is that advertising is simply an adjunct to merchandising, and serves to inform the public about products available. This is, of course, a profound fraud. Madison Avenue is the Defense Department of the society of corporations. It is the purveyor and enforcer of the wisdom of the tools. It designs and deploys weapons of counterinsurgency against the assertion of the interests of natural humanity. It has subsumed and subverted journalism and much of public education. Anyone who can afford to can buy its power, but everyone who does, aids in redefining the world in its terms, that is to say, the terms of the tools. Whoever utilizes Madison Avenue's capabilities is subsumed by them.

The power of corporate institutions to subsume more and more of the functions and institutions of non-technological aspects of society is illustrated in a series of events casually reported to the news media, regarding the ABM controversy.

When the Senate, made up of non-technicians and representing human as well as corporate constituencies, challenged the President on this weapons system, the President's staff called upon the potential recipients of contracts under the program to apply pressure (a public relations function) on the Senators from their States. This was worth one line and no comment in *Newsweek*. The President who does this was elected by virtue of a campaign directed by J. Walter Thompson, the world's largest advertising agency. The account supervisor from the Disneyland account directed the Presidential campaign for the agency, and became, after the election, the President's "chief of staff."

It is hard to escape the conclusion that the task of governing in the old sense of politics, values and human concerns, has been largely subsumed by and become subordinate to, the institutions managing the tools.

Thus the dilemma begins to take form. The Industrial Revolution has evolved into a system of tools and institutions which is absolutely necessary to the continued survival of populations and societies; and at the same time threatens mankind with sudden or gradual destruction. Eliminating the tools is not possible, without precipitating

114

the massive human die-off that perpetuating the present situation will also bring on. To place the tools under the control of the institutions that reflect human concerns above tool concerns, would seem to be to destroy the viability of the system, and precipitate the same dire result.

The gloomy circle depicted in these chapters is not closed or perfect. It is evolving. If it were perfectly developed, we no longer could discuss it. As Ellul points out, much of living has not yet been subsumed. But the process is frighteningly far advanced. And pollution of the public language is preparing the way.

EPILOGUE

Since this writing nearly forty years ago, many streams of events have flowed neatly from its premises. Of these I mention here two, which have seemed most clearly to illustrate the blind imperatives of technique, pursued by the great private firms, heedless of the potential for ultimate disaster. (A third, genetic engineering, may outstrip them all, if given time.)

The first of these is the SDI, or Strategic Defense Initiative. SDI is the current incarnation of a technological effort begun in 1945 (see p. 64). Under private control and public expense, it has included in sequence: surface-to-air weapons (Nikes); surface-to-space weapons (ABMs); and space-to-space weapons (Star Wars), with the implied intent to elaborate space-to-surface weapons. Scores of billions of dollars and centuries of time of highly skilled persons have been flushed into this septic drainpipe. "What can be done will be done." Fortunately none of the resulting systems has ever been shown to work.

The other stream of tool development seems to work all too well. Indeed the capability to create and integrate data banks, and to troll the world-ocean of microwaves for information from and about the earth's inhabitants, has no doubt exceeded our wildest nightmares, allowing for continuous surveillance of whole populations. That arbitrary use of this power is illegal seems of little moment. "What can be done will be done." A real danger now lurks that the fruit of centuries of struggle toward humane, democratic forms of social organization which include hard-won civil rights, may slide gently into this quicksand of total surveillance

These two historic streams of tool evolution share the characteristic of having been driven by carefully orchestrated fear, first of the Cold War, then of the fallout of the events of

September eleven. Each has resulted in veritable avalanches of public resources pouring into the tools of ultimate power.

I have been criticized for failing to suggest ways to escape from the witches' brew I have here attempted to describe. Sorry—I'm not up to it. It is no clearer to me today than it was thirty-eight years ago how the critical values of flesh-and-blood publics can be made to take precedence over the cold, narrowly legalistic rationality of global corporate interests. Indeed, with the elaboration of such entities as the World Trade Organization (WTO), the World Bank, and various "free trade" agreements, the prospect seems to be dimming. Added to these concerns is the equally ominous creeping privatizing of police and military functions.

With these and other traditional legitimate powers of governments to act in the public interests of their citizens being eroded, in favor of the commerce-driven interests of the global corporations, one might ask how the principle of product liability might be brought to bear on the effects of cruise missiles, work-performing robots, treatment-resistant pathogens, genetic engineering and greenhouse gases. At least, of the current order of things, we must relentlessly ask, "*Cui bono*—Who benefits?" until we are content with the answer.

BIBLIOGRAPHY

THE WISDOM OF MAN

Kahlil Gibran. *The Prophet*, Alfred Knopf, New York, 1943.
    (The best statement I have found or expect to find of the wisdom of Man; by which one can measure sanity in another wisdom.)

INSPIRING AUDACITIES

Robert Ardrey. *African Genesis*, Dell Publishing Co., 1961.
------------. *The Territorial Imperative*, Atheneum, NewYork, 1966.
Konrad Lorenz. *KingSolomon'sRing*, Cromwell, NewYork, 1952.
-----------. *On Aggression*, Harcourt, Brace & World, Inc., New York, 1966.
Pierre Teilhard de Chardin. *The Phenomenon of Man*, Harperand Row, New York and Evanston, 1961.
    (Without people like these to lead the way, who would have the temerity to attack our best beloved errors?)

THE ART OF FOREBODING

Marston Bates. *The Prevalence of People*, Charles Scribner's Sons, New York, 1956.
Georg Borgstrom. *The Hungry Planet*, The Macmillan Company, New York, 1965.
Nigel Calder (ed.). *Unless Peace Comes*, The Viking Press, New York, 1968.

Rachel Carson. *Silent Spring*, Fawcett World Library, New
York, 1962.
(Other books of foreboding are appendices and
memorials to this one.)
Stuart Chase. *The Most Probable World*, Harper and Row,
New York and Evanston, 1968.
Barry Commoner. *Science and Survival*, The Viking Press,
1963.
Paul Erhlich. *The Population Bomb*, Ballantine Books, New
York, 1968.
William and Paul Paddock. *Famine—1975*, Little, Brown
and Company, Boston and Toronto, 1967.
William Vogt. *Road to survival*, William Sloane Associates,
Inc., New York, 1948.
(For the sake of one's sanity, disposition and family,
these should be taken in small doses, like castor oil.)

THE TOOLS AND THEIR MASTERS

Jacques Ellul. *The Technological Society*, Alfred Knopf,
New York, 1967.
(He poses unanswerable questions, and does not
answer them; which is refreshing and most helpful.)
John Kenneth Galbraith. *The Affluent Society*, Houghton
Mifflin Company, Boston, 1958.
------------. *The New Industrial State*, Houghton Mifflin
Company, Boston, 1967.
(A proper portrait of the corporation. Another
inspiring audacity.)
Walton Hamilton. *The Politics of Industry*, Alfred Knopf,
New York, 1957.
(Hamilton has a cold and clinical eye for what is
going on.)
Anthony Jay. *Management and Machievelli*. Holt, Rinehart
and Winston, New York, 1967.

Edward S. Mason (ed.) *The Corporation in Modern Society*, Atheneum, New York, 1966. (Harvard University Press, 1959). (A gold pan full of nuggets.)

Seymour Melman. *Our Depleted Society*, Dell Publishing Company, New York, 1965.
(Scary; Chapter 9 belongs among the forebodings.)

C. Wright Mills. *The Power Elite*, Oxford University Press , New York, 1959.
(Forty-three is too young to die.)

Ralph Nader. *Unsafe At Any Speed*, Pocket Books, New York, 1965.

H.L. Nieburg. *In the Name of Science*, Quadrangle Books , Chicago, 1966.
(A wonderfully detailed scrutiny of the world of weapons industry, *vis a vis* the Department of Defense.)

Drew Pearson and Jack Anderson. *The Case Against Congress*, Simon and Schuster, New York, 1968.
(In these days of PR, one takes journalism where one finds it and rejoices.)

Robert Perrucci and Marc Piliauk. *The Triple Revolution*, Little, Brown and Company, Boston, 1968.

James Ridgeway. *The Closed Corporation*, Ballantine Books, Inc., New York, 1968.

J.-J. Servan-Schreiber. *The American Challenge*, Atheneum, New York, 1968.
Carl Van Doren. *The Great Rehearsal*, The Viking Press, New York, 1948.
(A delightful account of the making of the U.S. Constitution.)

Norman Wengert. *Natural Resources and the Political Struggle*, Random House, New York, 1955.

The Center for the Study of Democratic Institutions, Santa Barbara.
(The *Center Magazine* and the *Occasional Papers* have maintained a continuing dialog for ten years on topics relating to industrial society.)

THE LANGUAGE OF THE TOOLS

Samm Sinclair Baker. *The Permissible Lie*, The World
Publishing Company, Cleveland and New York,
1968.
('The Inside Truth About Advertising" doesn't take
one very far inside, but glimpses are interesting.)
Daniel Bell (ed.). *Toward the Year 2000*, Houghton Mifflin
Company, Boston, 1967.
Kenneth Burke. *The Philosophy of Literary Form*, Alfred
Knopf, New York, 1957. (Louisiana University
Press, 1941.)
Paul Goodman. *Growing Up Absurd*, V. Gollancz, London,
1960.
(The predicament of perceptive kids.)
Joint Council on Economic Education. *Economics and the
Consumer*, New York, 1966.
(This instructive little pamphlet illustrates what those
who ought to know would like the children to think is
going on.)
Kenneth Kenniston. *The Uncommitted*, Harcourt Brace &
World, Inc., New York, 1960.
Herbert Marcuse. *One-Dimensional Man*, Beacon Press,
Boston, 1964.
(The best single statement I have found describing
what is happening to the language of public
discourse.)
Marshall McLuhan. *Understanding Media*, McGraw-Hill
Book Company, New York, 1964.
Norbert Wiener. *The Human Use of Human Beings*,
Houghton Mifflin Company, Boston, 1950.